Acknowledgements
Manual H
Second Edition

This Manual was prepared by the
Air Conditioning Contractors of America

Author:

Hank Rutkowski, P.E.
ACCA Technical Director

Review and Technical Assistance:

Rod Beever
Product Manager, Borg-Warner Environmental Systems, Inc.,
(York, PA)

Don Cochell
Technical Advisor, Detroit Edison (Detroit, MI)

R. Lee Culpepper
Program Manager, Tennessee Valley Authority
(Chattanooga, TN)

Stephen D. Kennedy
Southern Company Services (Atlanta, GA)

Nance Lovvorn
Technical Advisor, Alabama Power Company
(Birmingham, AL)

Thomas McGarry
Residential Marketing, Training Director
Rochester Gas & Electric (Rochester, NY)

Barney Menditch, P.E.
President of General Heating Engineering Co., Inc.
(Capitol Heights, MD)

Alfred A. Piff
President of A/C Contracting Company, Inc.
(Mobile, AL)

Table of Contents

ISBN: 978-1-892765-07-9

List of Figures

List of Tables

List of Abbreviations

ACCA - Air Conditioning Contractors of America
NAHB - National Association of Home Builders
ASHRAE - American Society of Heating, Refrigerating, and Air Conditioning Engineers
ARI - Airconditioning & Refrigeration Institute
NFPA - National Fire Protection Association
FHA - Federal Housing Authority
HUD - Department of Housing and Urban Development
AFM - Air Force Manual
NWWA - National Water Well Association

List of Symbols

Btu - British Thermal Unit
Btuh - Btu Per Hour
CMF - Cubic Feet Per Minute
cu. ft. - Cubic Foot
D.D. - Degree Days
E - Efficiency of Equipment
ESP - External Static Pressure
°F - Degree Fahrenheit
GAL - Gallon
GPM - Gallons Per Minute
H - Convective Heat Transfer Coefficient
HL - Heating Load
IN WG - Inches of Water Gauge Pressure
KW - Kilowatt
KWH - Kilowatt Hour
MCF - 1,000 Cubic Feet
OAT - Outside Air Temperature
Q - Heat Flow - Btu/hr.
R - Thermal Resistance = sq.ft. x F x hr./Btu
RAT - Room Air Temperature
sq. ft. - Square Foot
T.D. - Temperature Difference F
TON - 12,000 Btu/hr.
V - Volts

Preface

It is assumed that the user of this manual is already well grounded in the basic refrigerant cycle and the knowledge and procedures (Manual J) needed to calculate residential heating and cooling loads. It is further assumed that the user is proficient in residential equipment selection and system design (Manuals J, D, & G). With this prerequisite knowledge and skill, the student of this manual will be able to develop comparable skill in heat pump equipment selection and system design quite easily.

The user of this manual is hereby cautioned to follow the requirements of applicable local, state, or federal statutes or codes as well as specific recommendations or requirements of manufacturers in any case where there may be disagreement with the statements in this manual.

Introduction

This manual includes information on all heat pump systems and equipment for both residential and small commercial applications. The manual is divided into five sections which are summarized below.

Basic Principles

Section 1 includes a discussion of the factors which generally apply to heat pump selection, performance and operation. These topics include integrated capacity, COP, theoretical and operating balance points, sizing considerations for various climates, comparison of air source and water source performance characteristics, operating cost, thermodynamic principles, auxiliary and emergency heat requirements.

Overview of the Various Types of Heat Pump Systems

Section 2 includes explanations of how the various types of heat pumps operate. This section also includes comments on performance, advantages and disadvantages, first cost, operating cost, and applications of the various types of systems.

Those systems discussed include air to air, water to air, ground water, air to water, water to water, add on, hybrid, dual fuel heat pumps, reclaim systems, zoned systems, earth coupled heat pumps, solar assisted heat pumps and heat pump water heaters.

Heat Pump Operation

Section 3 includes information on the major components used in heat pump systems. This section also includes information on refrigeration and reversed refrigeration cycles, suitability of heat sources and sinks, safety and operating controls, defrost cycles and controls, efficiency ratings, control strategies for the various types of heat pump equipment, and control cycles for reducing energy consumption.

Application and System Design

Section 4 outlines the design procedures required for both residential and commercial applications. Reference is made to load calculations, air terminal selection, duct design and water piping calculations. This section also includes examples of air source and water source equipment selection procedures and detailed examples of residential energy consumption calculations.

Installation Procedures

Section 5 provides a quick review of all factors which should be considered when the equipment is actually installed. Topics include equipment location, accessability, refrigerant piping, refrigerant charge, controls, and power wiring.

Section 1

General Discussion of Heat Pump Basics

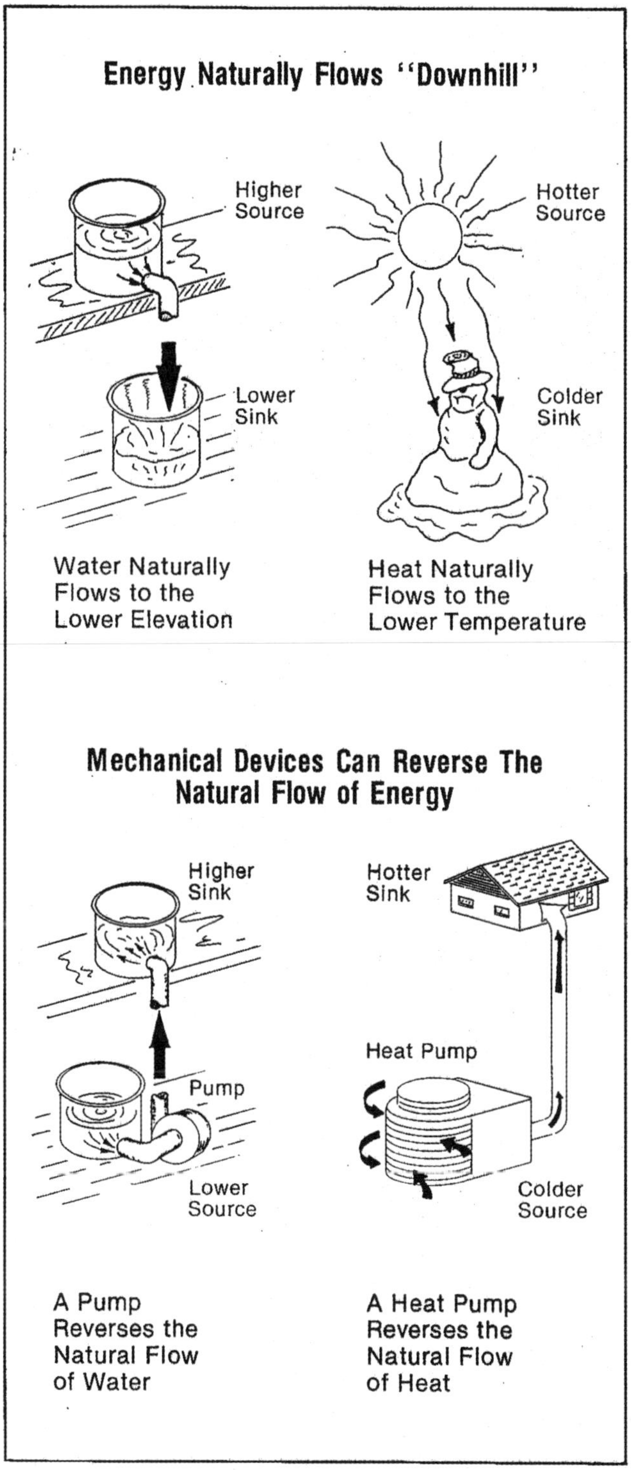

Figure 1.0 Energy Flow

From the laws of nature, we know that water will always flow from a higher elevation to a lower elevation unless a mechanical device is used to reverse the flow. Normally this device is a pump which works to overcome gravity by making the water flow uphill.

Unless a mechanical device is used to reverse the flow, heat will always flow from the higher temperature "source" to the lower temperature "sink". However, if a refrigeration machine (heat pump) is used, heat can be made to flow in the reverse direction. These machines "pump" heat from a "source" at a lower temperature to a "sink" at a higher temperature. Figure 1.0 illustrates how the water pump and the heat pump are used to reverse the natural flow of energy.

We usually associate the term "heat pump" with a heating device, but cooling equipment also functions as a heat pump. For example, a refrigerator is a heat pump because it moves heat from a cold inside space to the warmer exterior space. Likewise, the room air conditioner is a heat pump because it makes heat flow from a lower temperature area to a higher temperature area. The terms "refrigerator" and "air conditioner" are commonly associated with cooling equipment. These devices are not normally called "heat pumps", but when a refrigeration machine is used for heating, it is usually called a "heat pump". Heat pumps are usually classified by the types of sources and sinks that are used with the heat pump. These sources and sinks are usually fluids such as air or water. The diagrams shown in Figure 1.1 illustrate the different types of heat pump equipment.

Air to Air Heat Pump

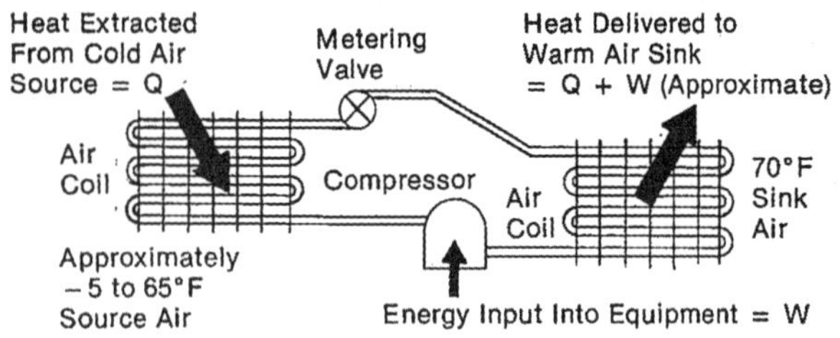

Water to Air Heat Pump

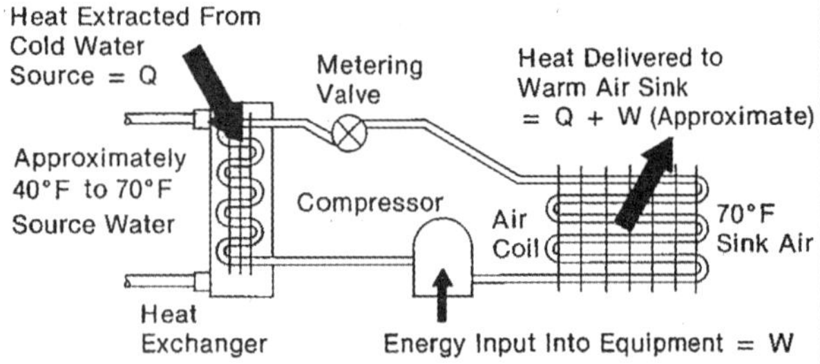

Air to Water Heat Pump

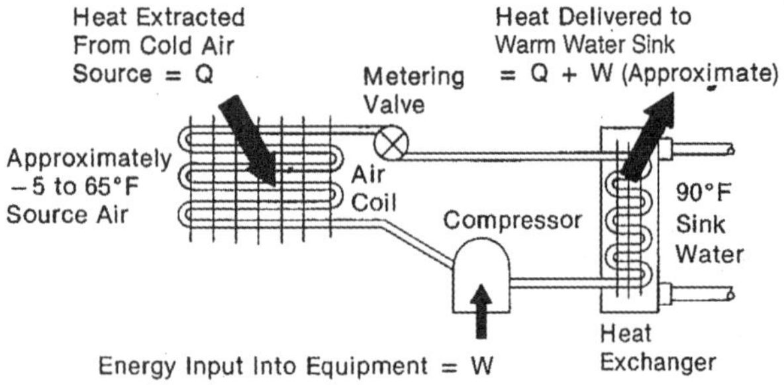

Water to Water Heat Pump

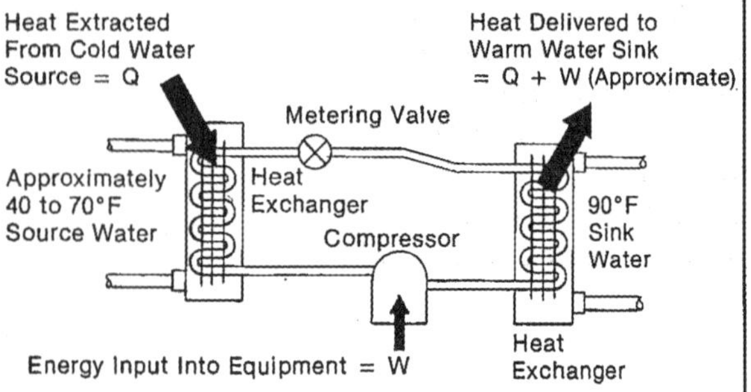

Figure 1.1 Different Types of Heat Pump Equipment

One important thing to remember is that heat pumps are devices which move heat. They do consume energy when they operate, but this energy is not directly used to produce heat. The energy consumed by the heat pump is used to move heat from a colder source to a warmer sink. Since energy is conserved, the energy which flows into the machine must equal the energy which flows out of the machine. To say it another way, the heat removed from the source plus the energy supplied to the machine must equal the heat rejected to the sink. It is the heat pump's ability to move considerably more energy than it consumes that makes it so effective as a heating device.

Figure 1.2 shows that the energy flow into and out of the refrigeration cycle are equal. Heat pumps also require energy for fans or pumps (auxiliaries). Figure 1.2 also shows that the energy flow into the heat pump system is equal to the energy flow out of the heat pump system.

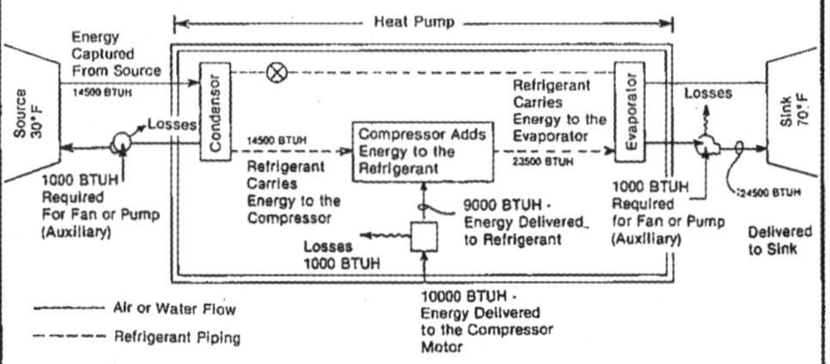

Energy Into the Refrigeration Cycle

Energy From Source	=	14500
Energy Delivered to the Compressor	=	10000
		24500 BTUH

Energy Out of the Refrigeration Cycle

Energy Delivered to the Evaporator	=	23500
Refrigeration Cycle Losses	=	1000
		24500 BTUH

Total Energy Into the System

Energy Captured From the Source	=	14500
Energy Delivered to the Compressor	=	10000
Energy for Auxiliaries	=	2000
		26500 BTUH

Total Energy Out of the System

Energy Delivered to the Sink	=	24500
Refrigeration Cycle Losses	=	1000
Auxiliary Losses	=	1000
		26500 BTUH

Figure 1.2 Energy Flow in the Refrigeration Cycle

1.1 Heat Extraction

The idea of using cold outdoor air to heat a house is a little confusing. But, the terms "cold" and "hot" are relative, and the laws of physics show that any substance contains heat if it has a temperature above "absolute zero." Since absolute zero is approximately equal to 460 degrees below zero (-460°F) cold winter air or the cool ground water still contains an ample amount of heat which can be recovered by a heat pump. The exact details of how the heat pump does this will be discussed in another section. For now it is only necessary to understand that the heat pump is able to extract heat from relatively cold sources. Figure 1.3 compares the amount of heat that is available in one pound of air or water at various temperatures.

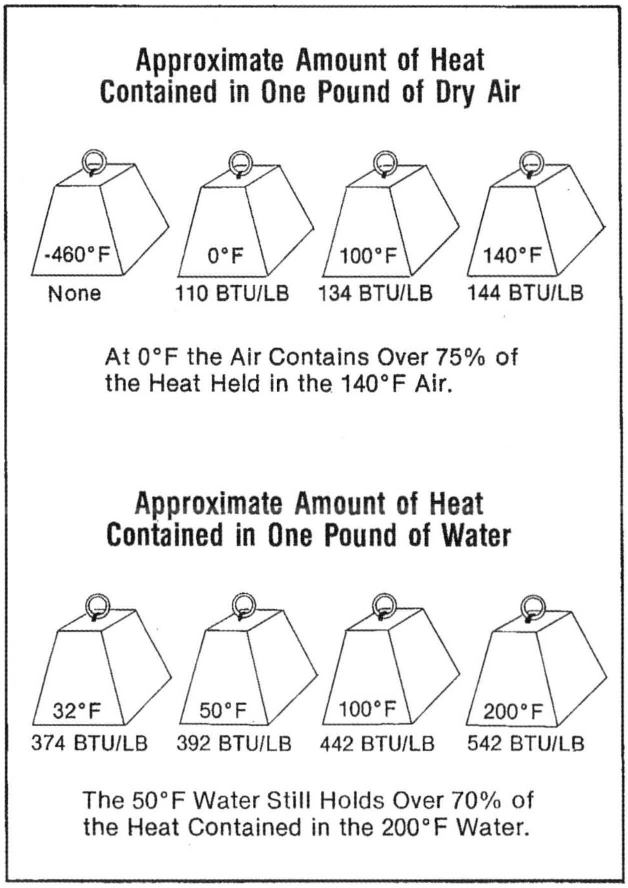

Figure 1.3

1.2 Heat Pump Performance Characteristics

Heat pumps extract heat from cold sources, but this becomes more difficult as the source temperature drops. Because the output of the heat pump depends on the temperature of the source, the heat pump performance must be shown by a graph which correlates the heat output with the source temperature. The performance curves illustrated by Figure 1.4 illustrate how the source temperature affects heat pump capacity. In the figure, the performance of the air source heat pump is compared to the performance of a ground water heat pump. Note how the ground water heat pump capacity remains relatively constant at any outdoor temperature while the air source unit experiences a loss of capacity as the outdoor temperature falls. This is because the temperature of the ground water remains relatively constant and is only slightly affected by outdoor temperature changes.

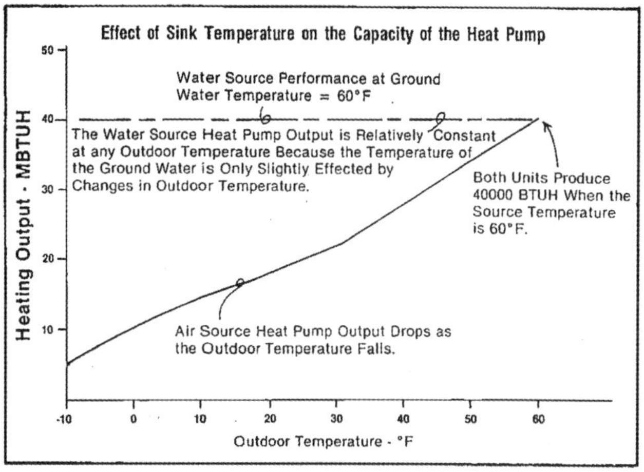

Figure 1.4

1.3 Integrated Capacity

Air source heat pumps tend to build up frost or ice on the outdoor coil when the outdoor temperature falls below 45°F. Since this is undesirable, these machines are equipped with devices which automatically defrost the coil. When these devices are in operation, they consume energy and reduce the heat pump efficiency. The "integrated capacity" includes allowances for the defrost cycle losses. Figure 1.5 illustrates how the defrost cycle affects heat pump performance. Always use the integrated capacity information when sizing air source heat pumps that operate in ambient temperatures below 45°F.

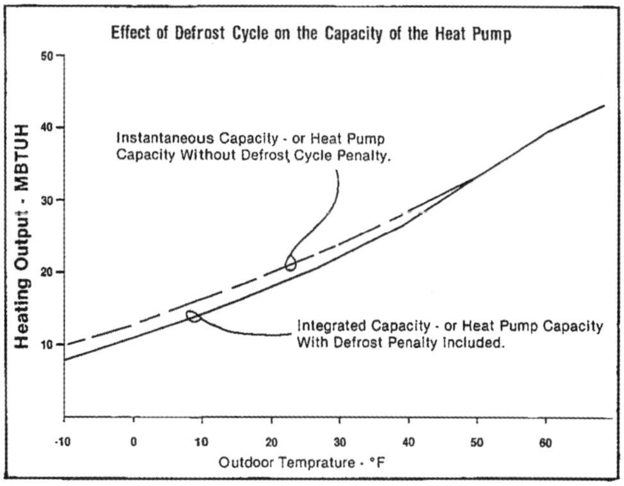

Figure 1.5

1.4 Coefficient of Performance

The ratio of the heat delivered by the heat pump to the energy required to operate the heat pump is called the coefficient of performance (COP).

$$\text{COP} = \frac{\text{Heat delivered (Btuh)}}{\text{Energy supplied to machine (Btuh)}}$$

The COP can be a misleading indication of efficiency because the COP of a heat pump varies with changes in the source and sink temperatures and is affected by defrost cycles and by the energy required for auxiliary devices. Figure 1.6 illustrates how the COP varies when the power required to operate the heat pump and the heat pump output are affected by the temperature of the source. Refer to Section 3.9 for more information on COP ratings.

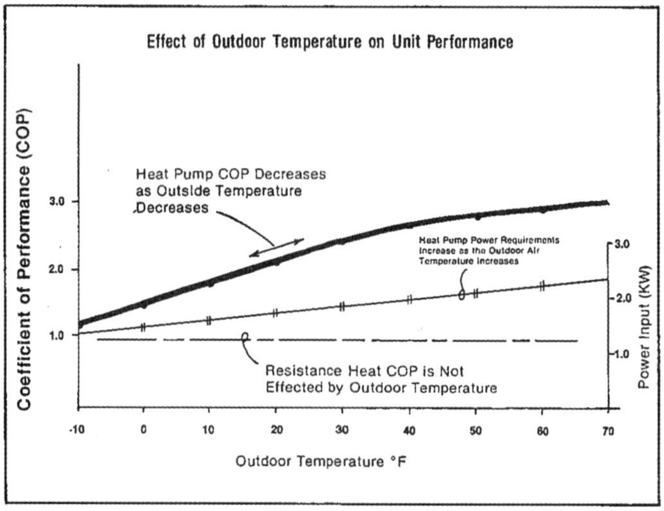

Figure 1.6

1.5 Heat Pump Sizing Considerations

If heat pumps are used for a "heating only" application, the design capacity can be equal to the winter design heating load. In this case, the heat pump will provide all the heat required by the building and no supplemental heat will be necessary. However, some manufacturers recommend sizing the unit at 75 percent to 85 percent of the winter design load with the remaining capacity made up by supplemental heat. This recommendation is based on the fact that the outdoor temperature will be above the winter design temperature for 97.5 percent of the annual heating hours. During these hours a slightly undersized unit will have adequate capacity and be more efficient.

Heat pumps are commonly used for both heating and cooling. In this case, the designer must attempt to optimize the heating and cooling performance of the unit. To do this correctly, the relative magnitude of the heating and cooling loads must be considered.

In colder regions, or in any situation where the heating loads are much larger than the cooling loads, the heat pump should not be sized for the design heating load. This is because a unit selection based on the design heating requirement will be oversized for cooling, and the cooling performance will be poor. A compromise must be made so that the selected unit will perform well in both the heating and cooling modes. In this case, the unit selection is normally based on the cooling requirement. However, it is permissible to oversize the unit for cooling by as much as twenty five percent to take full advantage of the heat pumps coefficient of performance during the heating mode. The objective is to select a unit that provides the most heating capacity without seriously degrading the cooling performance. Since the heat pump will not provide all the heat required by the structure on a winter design day, supplemental heat will be necessary.

In less severe climates, the design heating and cooling loads will be closer together. A unit that is sized so that it does not exceed the required cooling capacity by more than 25 percent may satisfy the design heating load without using supplemental heat.

In very warm regions, the design cooling load may equal or exceed the design heating load. Therefore, the unit heating capacity will be more than adequate if the unit is selected for the design cooling load.

Figure 1.7 illustrates the how the relative size of the design heating and cooling loads affect the unit selection.

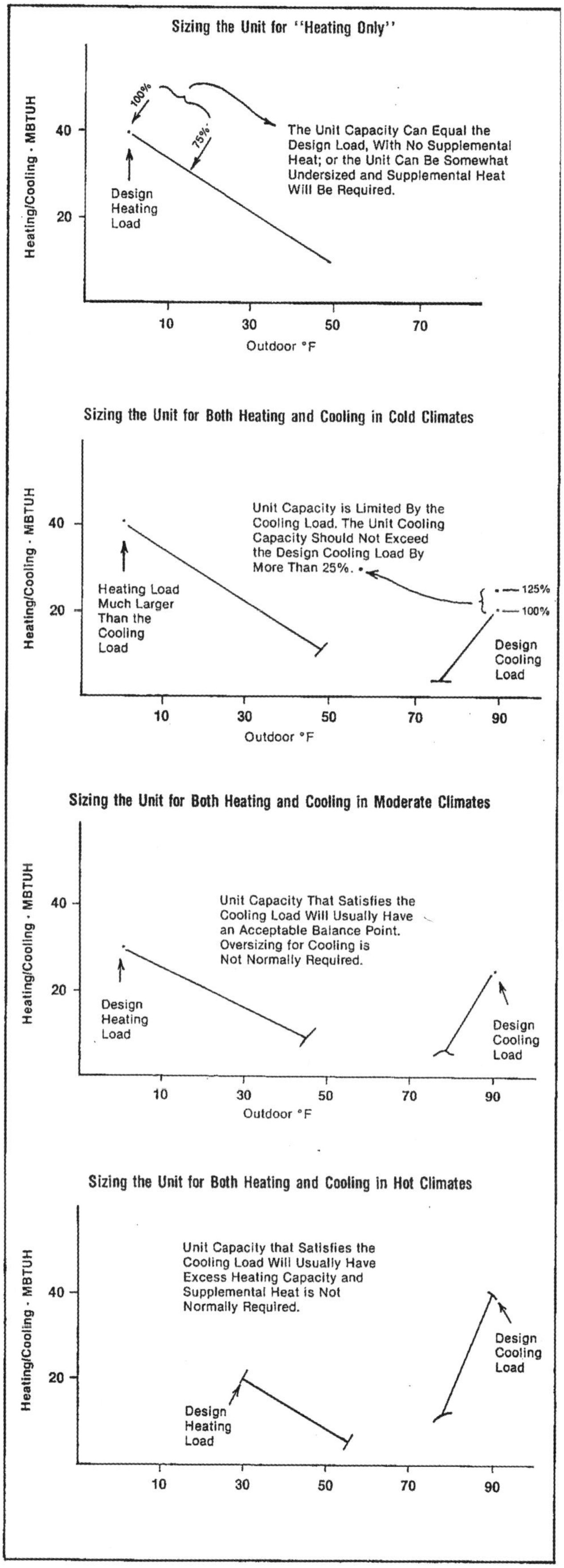

Figure 1.7 Size of Loads Affect Unit Selection

1.6 The Design Balance Point

The design balance point is a theoretical point on a graph which compares the heating requirements of the structure with the heat output of the heat pump at various outside temperatures.

For residential applications, the building heat loss is assumed to vary linearly with outside temperature. It is also assumed that in a typical residence, with the thermostat set at 70°F, heat is not required when the outdoor temperature rises above 60°F or 65°F. (There is controversy about which outside temperature is the most accurate "no load" temperature. In this manual, 65°F will be used as the "no load" temperature.) The building heat requirements can be estimated for any outside air temperature by drawing a straight line from the winter design condition to the "no load" condition. Figure 1.8 illustrates how this line is drawn.

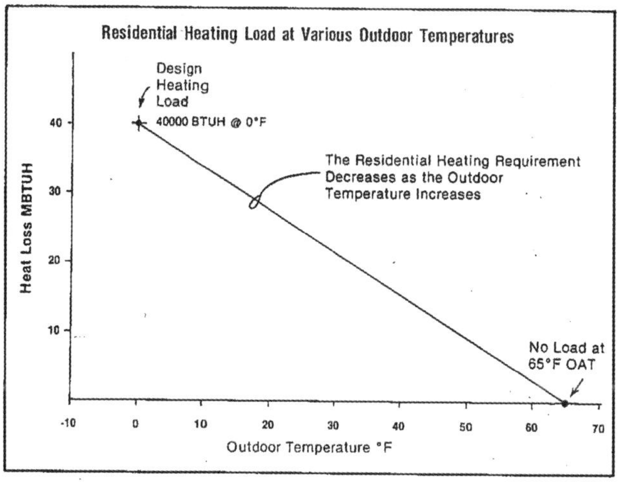

Figure 1.8

The heating output of the heat pump also varies with the outside air temperature as indicated in Section 1.3. If the heat pump performance line is superimposed on the building heating load line, the theoretical design balance point will be located at the intersection of the two lines. The outside temperature at which this occurs is the only temperature where the output of the heat pump exactly equals the building load. At temperatures above the balance point, the heat pump capacity will exceed the building load and the machine will cycle on and off under control of the room thermostat. At temperatures below the balance point, the heat pump will not be able to satisfy the room thermostat and supplemental heat will be necessary. Figure 1.9 illustrates how the design balance point is determined.

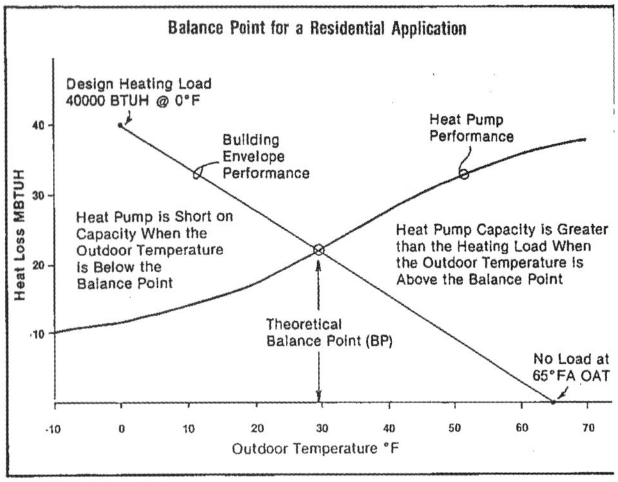

Figure 1.9

It is desirable to select a unit which has the lowest possible balance point without being oversized for cooling. A balance point of 30°F or lower is considered to be good, but the design cooling load may affect the selection of the design balance point. (Refer to Section 1.5 for a discussion on how the cooling load can limit the design heating capacity.) Note that it is also necessary to check local utility requirements. In some areas, the utility may stipulate a maximum allowable balance point. Figure 1.10 illustrates how the design balance point is determined when the cooling load is a limiting consideration.

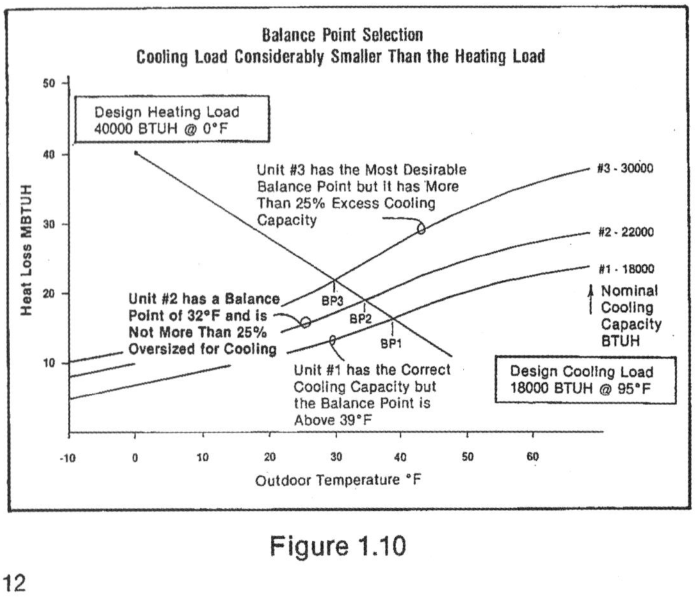

Figure 1.10

When the design cooling load is greater than the design heating load, the design balance point will occur at an outdoor temperature that is lower than the winter design temperature. Figure 1.11 illustrates this case.

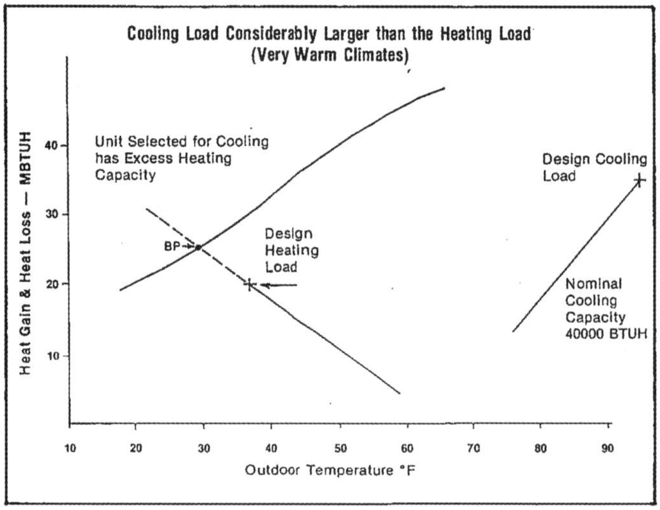

Figure 1.11

When heat pumps are used for commercial applications, the concept of determining the balance point is still valid. However, the method used for determining the heating and cooling loads at various outside temperatures is considerably different from the residential procedure described above. Each commercial structure will have a unique set of load conditions which must be evaluated on an individual basis. Consider the following:

During occupancy, commercial buildings can have large internal and/or solar gains. In this case, the "no load" condition for the space is usually different from the 65 degrees which was assumed for residential applications. In addition, mechanical ventilation produces loads on the unit which were not included in the residential procedure and which must be added to the space load.

When they are unoccupied, commercial buildings may not experience the large internal loads that occurred during the occupied hours. In this case, the "no load" condition in the space will be approximately equal to the night setback temperature. Also, the outside air dampers will usually be closed when a commercial building is unoccupied and the load on the equipment will be approximately equal to load in the space. (If the dampers are closed it is not unusual to find that they have significant leakage which produces some additional load on the equipment.)

Figure 1.12 illustrates how the loads in the perimeter and the interior zones of commercial structures vary with outdoor temperature for both the occupied and unoccupied conditions.

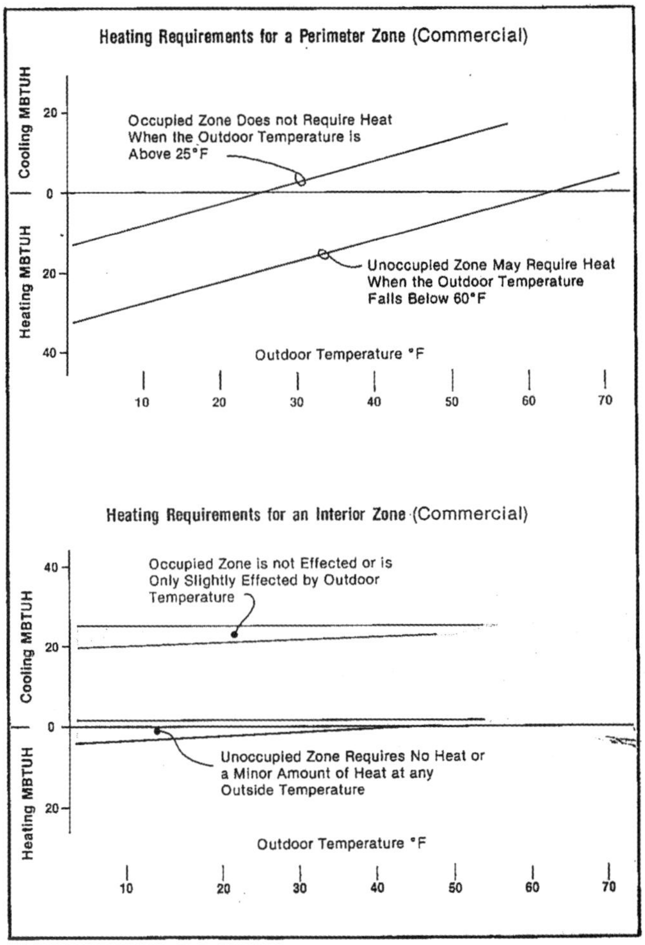

Figure 1.12 Loads in Perimeter and Interior Zones of Commercial Structures

1.7 Actual Operating Balance Point

When the heat pump is operating in the field, the actual balance point is seldom equal to the theoretical design balance point. This is because the design balance point was predicted by using a load calculation that assumed one specific value for the internal loads and because no allowance was made for the possibility of solar gains. It follows that the building load line and the balance point will vary as the internal loads and solar gains change. Other balance point variations occur because the actual equipment performance will not be exactly the same as the performance shown by the catalog data. Installation procedures, maintenance and local weather patterns are the source of the differences between the catalog performance data and the actual performance.

1.8 Supplemental Heat

When the outside air temperature is below the balance point, an additional source of heat is required to make up for the difference between the building load and the output of the heat pump. The amount of supplemental heat required is the difference between the heat loss of the structure and the heat pump capacity at the winter design temperature. As the outside temperature rises, this difference becomes less and less. No supplemental heat is required when the outside temperature is equal to or above the balance point.

Figure 1.13 illustrates how the loads on a commercial unit will vary with outdoor temperature for the occupied and unoccupied conditions.

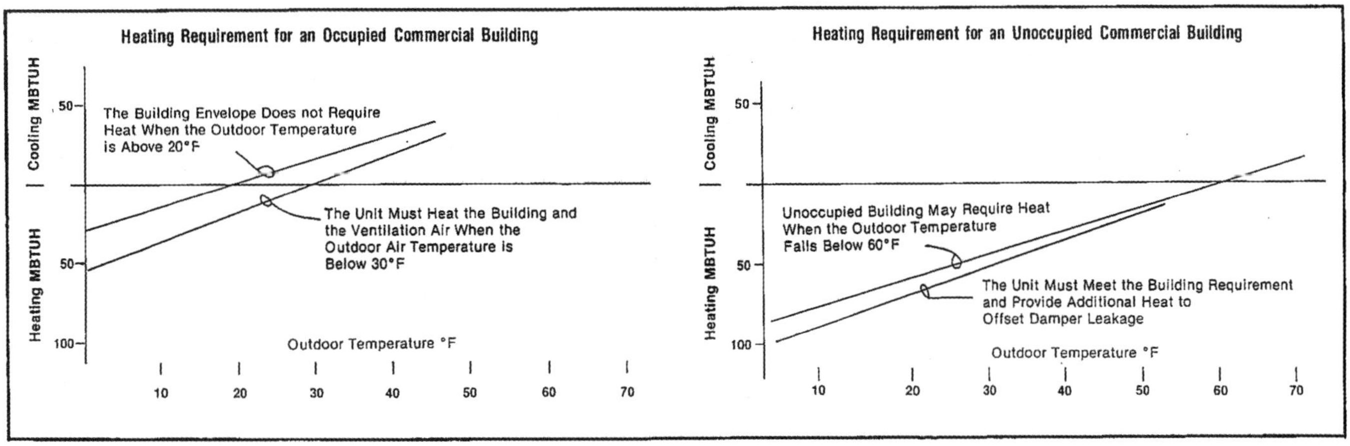

Figure 1.13 Loads on a Commercial Unit Vary with Outdoor Temperatures

Figure 1.14 illustrates how the supplemental heating capacity is determined from the balance point diagram.

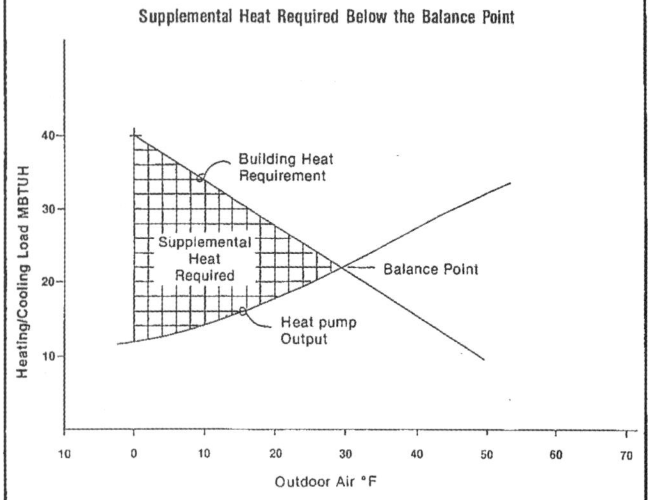

Figure 1.14

Normally, an electric heat pump is supplemented by electric resistance heaters. But, gas, oil, or propane furnaces can also serve in this capacity. However, fossil fuel backup is usually used when a heat pump is added to an existing fossil fuel heating system.

If electric resistance is used to provide the supplemental heat, it is economical to minimize the amount of energy which must be provided by the supplemental heating system. This is accomplished by sizing the heat pump for the lowest practical balance point and by continuous operation of the heat pump when the outdoor temperature is below the balance point.

If fossil fuels are used to provide supplemental heat, it may or may not be more economical to operate the heat pump at the lower outside temperatures. There is an "economic balance point" which determines the outside temperature at which the supplemental heat is less expensive than the heat supplied by the heat pump. This balance point depends on the cost of the fossil fuel, the cost of electricity, the COP of the heat pump and the efficiency of the supplemental heating system. (The economic balance point calculation is discussed in Section 2.5) .

Note that there are cases where it may be more economical to continue to operate the heat pump while the fossil fuel system is in operation, but this is not usually recommended. (Refer to Section 3.10 for a discussion of this subject.)

Figure 1.15 illustrates how equipment efficiencies and raw energy costs affect the cost of the heat that is finally delivered to the structure. Note that this figure is intended to illustrate how to evaluate the relative costs of energy delivered to the structure and it should not be used to make generalizations on the merits of the various fuels. In actual practice, the local fuel costs and the efficiencies of specific pieces of equipment must be used to evaluate the relative cost of the heat delivered to the space.

Comparison of Fuel Costs

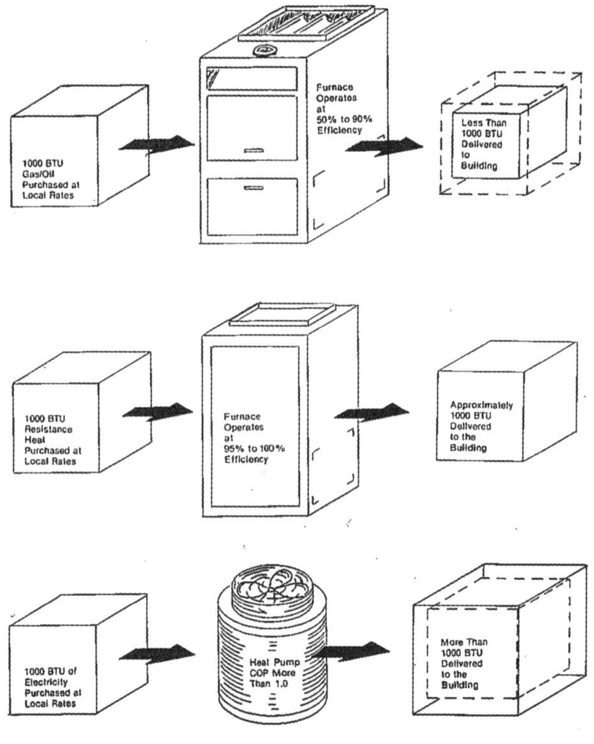

Example Fuel Costs @ 20°F Outdoor Temperature

Fuel	Cost For 1-MM BTU Raw	Efficiency or COP	Usable Heat (BTU)	Cost For 1-MM BTU Usable
Gas	$ 5.00	0.70	700,000	$ 7.14
Oil	$ 9.00	0.60	600,000	$15.00
Resistance	$14.65	1.00	1,000,000	$14.65
Heat Pump	$14.65	2.00	2,000,000	$ 7.32

***Costs Vary With Location and Time, Efficiency Varies With Equipment. — (1-MM Equal 1000000 BTU)

Figure 1.15

1.9 Emergency Heat

A compressor failure or other component failure may mean that the heat pump will be out of operation for an extended period. For this reason, some utilities or building codes may require that the capacity of the back up heating system be sufficient to maintain some minimum room temperature when the heat pump is out of operation Check local utility or code requirements to determine if emergency heat is required.

If emergency heat is required, the installed capacity of the supplemental system may be fixed by the "emergency heat" requirement. In this case, the supplemental system will be oversized when the heat pump is operating normally. If electric resistance heaters are used, they can have a high power draw and this can result in a short heating cycle. Over-sizing can be corrected by circuiting the resistance heaters so that only the supplemental KW is energized during normal operation but with a provision for energizing the additional KW for emergency operation. Figure 1.16 illustrates the difference between 100 percent emergency heat and the amount of supplemental heat which would be required for normal operation.

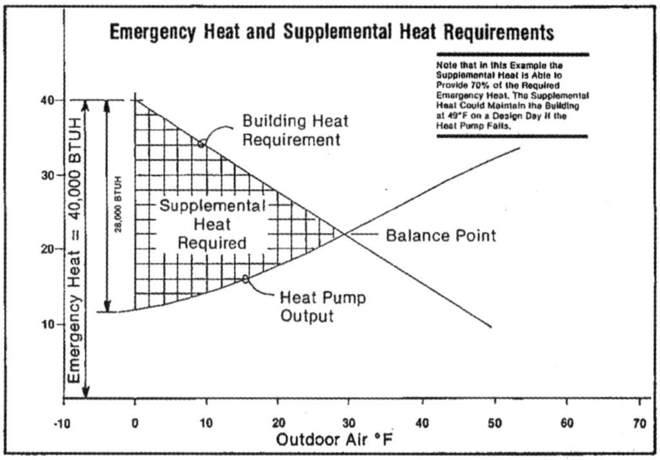

Figure 1.16

1.10 Operating Hours

Even in relatively cold climates, the heat pump will operate without supplemental heat a majority of the time and full supplemental capacity will only be required for a few hours per year. The result is that the supplemental heating system will only have to provide a minor share of the total yearly energy requirement. Cold climates require more supplemental energy than warmer climates. But, even in northern locations the relative amount of supplemental energy is minor. Figure 1.17 illustrates that in Green Bay, Wisconsin, only 20 percent of the annual heating energy is provided by the auxiliary heating system.

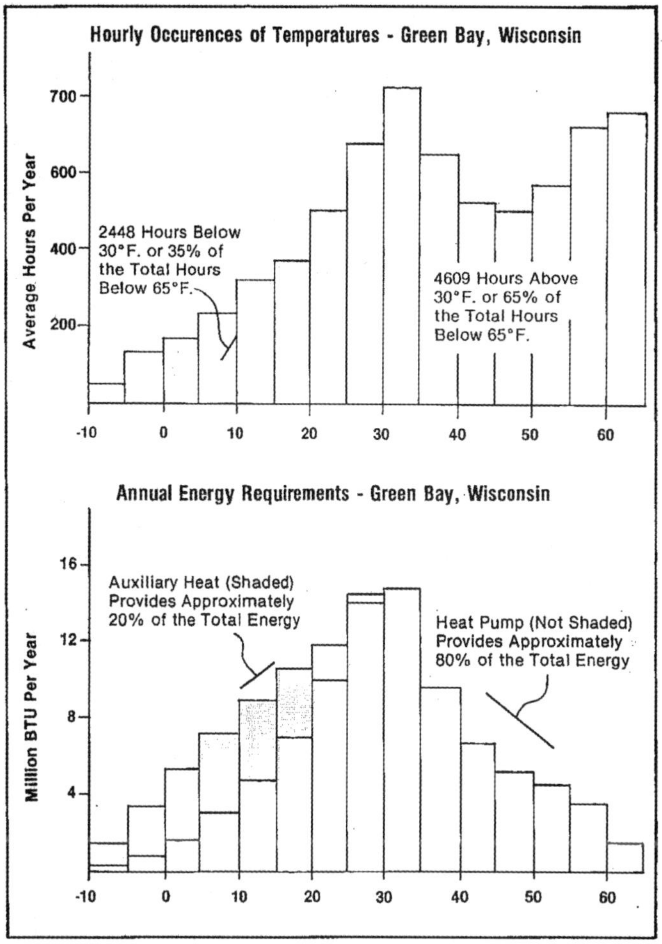

Figure 1.17

Section 2

Overview of Heat Pump Systems

The various types of heat pump systems are classified by the source from which the heat is extracted and the fluid which is used to distribute the heat to the occupied space. For example, the common residential heat pump is an air to air heat pump because it extracts heat from the atmosphere (air) and delivers this heat to the air that is circulated through the structure. However, this method of classification is not always used when the various devices are marketed. Terms such as "geothermal", "earth-coupled", "water-source", "add-on", "dual fuel", or "solar-assisted", do not completely describe the device. In this section, all the common (and some that are not so common) heat pump systems will be identified and discussed.

Heat pumps are most often identified with their operation in the heating mode, but it should be emphasized that heat pumps can also be used to provide summer cooling. When the equipment cycle is reversed to provide cooling, the source and sink are also reversed. For example, a home that is heated with a water to air heat pump in the winter will be cooled by an air to water heat pump in the summer. But, the equipment will be referred to as a water to air heat pump because it operates in this mode when it is used as a heating device.

The physical arrangement of the various components is an additional consideration when heat pump systems are classified. When all the components are housed in one enclosure, the equipment is referred to as a "single package system."

This type of unit must be installed so that both the source fluid and sink fluid are available at the unit. Single package equipment may be designed to be used with a supply and return duct system (unitary heat pump), or it may be designed for through the wall installation (packaged terminal heat pump).

"Split systems" divide the various components into separate units. The indoor unit contains the indoor heat exchanger and the outdoor unit contains the compressor and the external heat exchanger. Sometimes the compressor is supplied separately in a third unit which can be installed inside with only the outdoor heat exchanger remaining in the outdoor unit. The split system allows the various components to be installed where access to the source and sink fluids is most convenient, but these units must be connected with field installed refrigerant piping and control wiring.

Figure 2.0 illustrates the various physical arrangements as they are applied to the common air to air heat pump.

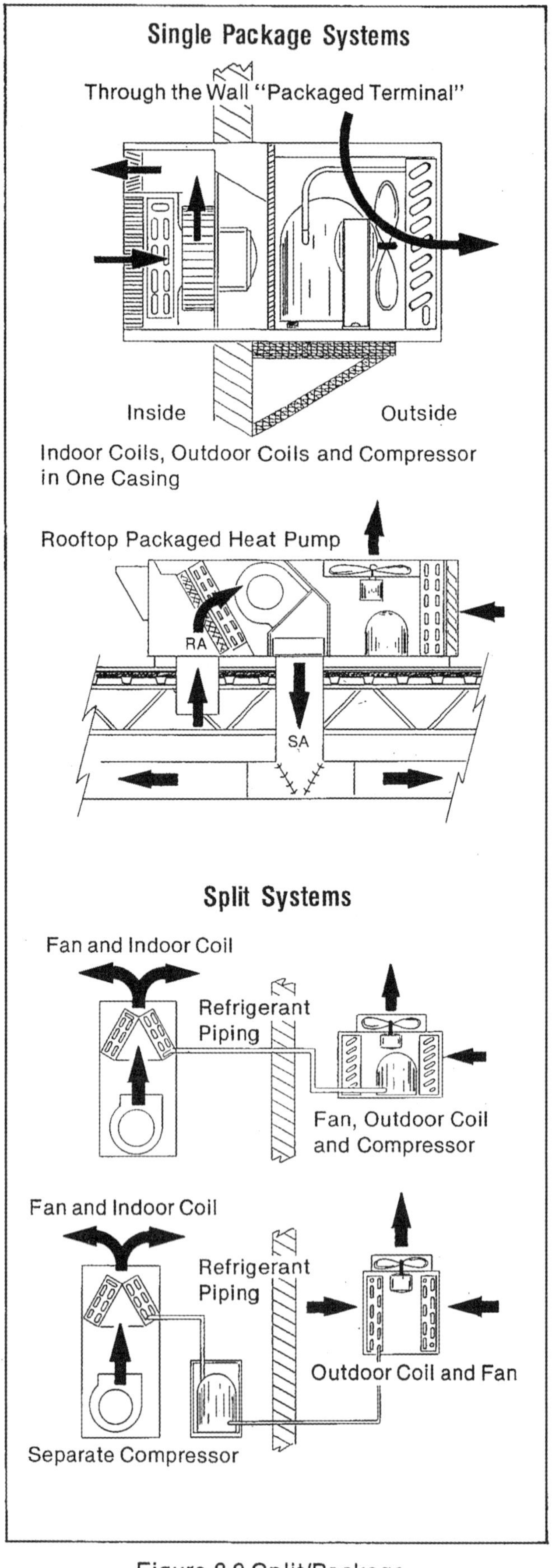

Figure 2.0 Split/Package

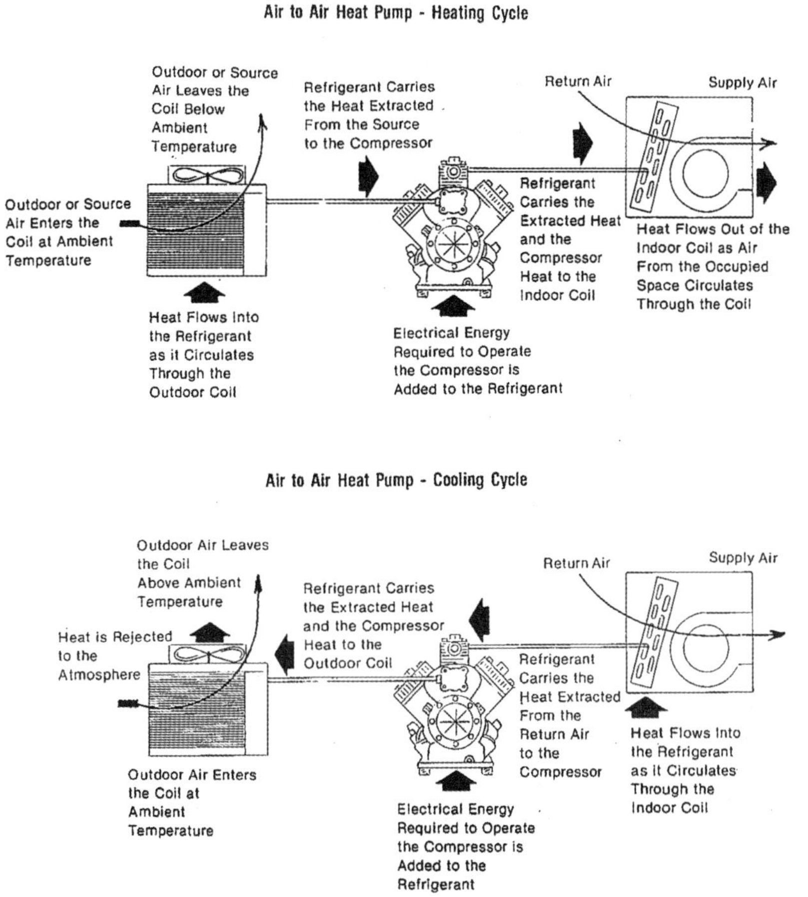

Figure 2.1 Air/Air Energy Flow

2.1 Air to Air Heat Pumps

These devices extract heat from one air stream (the source) and deliver the heat to the air stream that is circulated through the occupied space (the sink). The most common applications use the atmosphere as a heat source, but these devices can also be used to recover heat from air that would otherwise be exhausted from a building or process. Figure 2.1 illustrates the energy flow for the air to air heat pump heating and cooling cycles.

Air to air equipment is the most common heat pump system because the atmosphere is so readily available as a heat source. The air to air heat pump is used for residential, commercial and industrial applications and is available as a single package or split system. "Through the wall" equipment which slides into a pre-installed wall sleeve is available for the apartment and condominium market. "Zoned" heat pump systems are available for both residential and commercial applications. There is no specific upper limit on size, but larger units are usually fifty tons or less. Small equipment is available in sizes of less than one ton.

The air to air heat pump can also be used to recover heat from building or process exhaust systems. In this application, the air to air heat pump will operate at a high COP because the source temperatures are usually 50°F or higher. Another advantage in using the air to air heat pump for this application is that this equipment allows the recovered heat to be delivered to a location that is remote from the source air stream.

One drawback associated with the air to air heat pump is that the efficiency (COP) of this equipment is reduced as the source air temperature becomes colder. Fortunately, the average winter temperatures in most regions of North America are high enough to allow this equipment to operate at a desirable COP. However, in colder climates a second source of heat is required to supplement the heat pump when the outdoor temperature drops below the balance point. Another difficulty associated with the air to air heat pump arises because frost can form on the outdoor coil when the outdoor temperature falls below 45°F. This problem has been overcome by the development of reliable defrost devices. However, the defrost cycle does have an effect on heat pump efficiency because the energy that is used to defrost the coil is not available to heat the structure.

Installation cost for air to air systems depends primarily on the equipment size, air distribution system, and control options. Electrical resistance coils are usually used as supplementary heat on new installations. Since ambient air is readily available, there is no additional cost associated with developing the heat source.

The operating cost of air to air heat pumps is usually lower than, or competitive with, other types of fossil fuel and heat pump equipment. When operating cost comparisons are made, the heating and cooling requirements of structure, local weather patterns, local electric rates and fossil fuel costs, furnace efficiencies and/or the performance of alternate heat pump equipment must be considered. Realistic comparisons can only be made for specific pieces of equipment installed in a given geographic area. However, certain generalizations can be made:

A) The cost of heat supplied by the all electric heat pump will always be less than the cost of an equivalent amount of electric resistance heat.

B) The operating cost of water source or solar assisted heat pumps are normally lower than air source equipment operating cost, but usually have a substantially higher first cost.

C) An air source heat pump with fossil fuel supplementary heat can be less expensive to operate than an all electric heat pump. However, this depends on the relative cost of electricity, fossil fuel, and the operating efficiencies of the equipment.

2.2 Water to Air Heat Pumps

Water to air heat pumps extract heat from a water source and deliver heat to the indoor supply air stream. Usually a well, pond or lake provides the source water. (Since well water temperatures are not greatly affected by seasonal changes, wells provide consistent performance during all seasons. However, ponds and lakes can be subject to large temperature swings which can adversly affect equipment performance.)

This equipment can also extract heat from other fluid sources and it is especially versatile when used for heat recovery applications. It can efficiently recover heat from water, or other fluids which are discharged from building water systems or manufacturing processes.

The water to air heat pump can also be used to recover heat that is generated within the building envelope. This is accomplished by using a closed water loop to transfer heat from zone to zone within a building. When two or more of these units are connected to a common water piping loop, they are able to extract heat from the air in one area of a building where cooling is required, and deliver that heat to another area where heating is required.

The water source heat pump can also be used to extract heat from a water loop heated by solar energy. This arrangement increases the heating ability of water that circulates through the solar collectors.

2.3 Ground Water Heat Pumps

Ground water heat pumps can be used for residential, commercial and industrial applications and they are available in sizes that range from several tons to over fifty tons. This equipment can be used wherever an adequate supply of water is available. In most cases, the ground or surface water water temperature must be 40°F or higher for heating. (Always refer to the manufacturers recommendations for the temperature limits that apply to a particular piece of equipment.)

The map shown in Figure 2.2 illustrates the availability and temperature of the major aquifers located throughout the United States.

Figure 2.2 Map of Aquifers

Ground water equipment is manufactured in two distinct configurations. Both designs operate in essentially the same manner when they are heating, but they operate quite differently in the cooling mode. One type of system uses the compressor to provide mechanical cooling, while the other design provides summer cooling by circulating the ground water directly through a separate water coil located in the air handler. Note that the ground water heat pump size will be affected by the method that is used to provide cooling.

Mechanical Cooling. In this configuration, the refrigeration cycle is reversed, and the equipment extracts heat from the air in the occupied space and delivers this heat to the ground water. In this mode, the equipment operates as an air conditioner with a water cooled condenser. (The ground or surface water temperature must be 90°F or lower.) If the heating load is much larger than the cooling load, the equipment will have to be sized to satisfy the design sensible and latent cooling loads. In this case, supplemental heat will be required below the balance point. The considerations involved are essentially the same as those which are outlined in Section 1.5 of this manual.

Ground Water Cooling. In this configuration, the equipment is designed to provide cooling by circulating the ground water directly through a water coil located in the air handler. The compressor is not used for cooling, and the equipment can be selected for "heating only" (refer to Section 1.5). The water coil used for cooling should be selected to satisy the design sensible and latent cooling loads. Water coil cooling usually provides ample sensible cooling but the latent capacity is limited by the supply water temperature. Water temperatures that fall between 45°F and 50°F are normally required for good dehumidification, but applications that have small latent loads may be satisfied by water temperatures that are somewhat higher.

Figure 2.3 illustrates the energy flow for typical ground water heating applications.

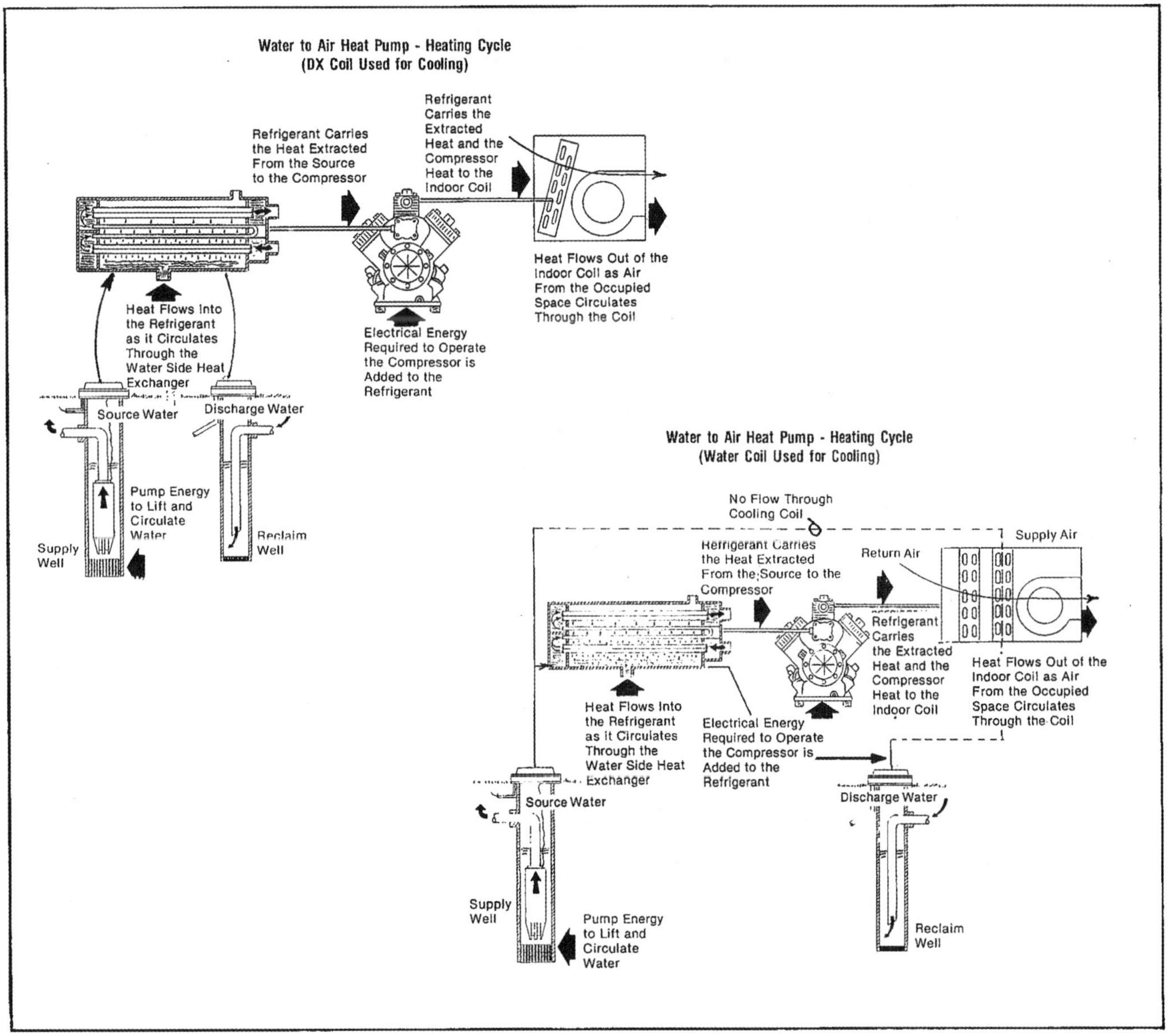

Figure 2.3 Heating Cycle Energy Flow

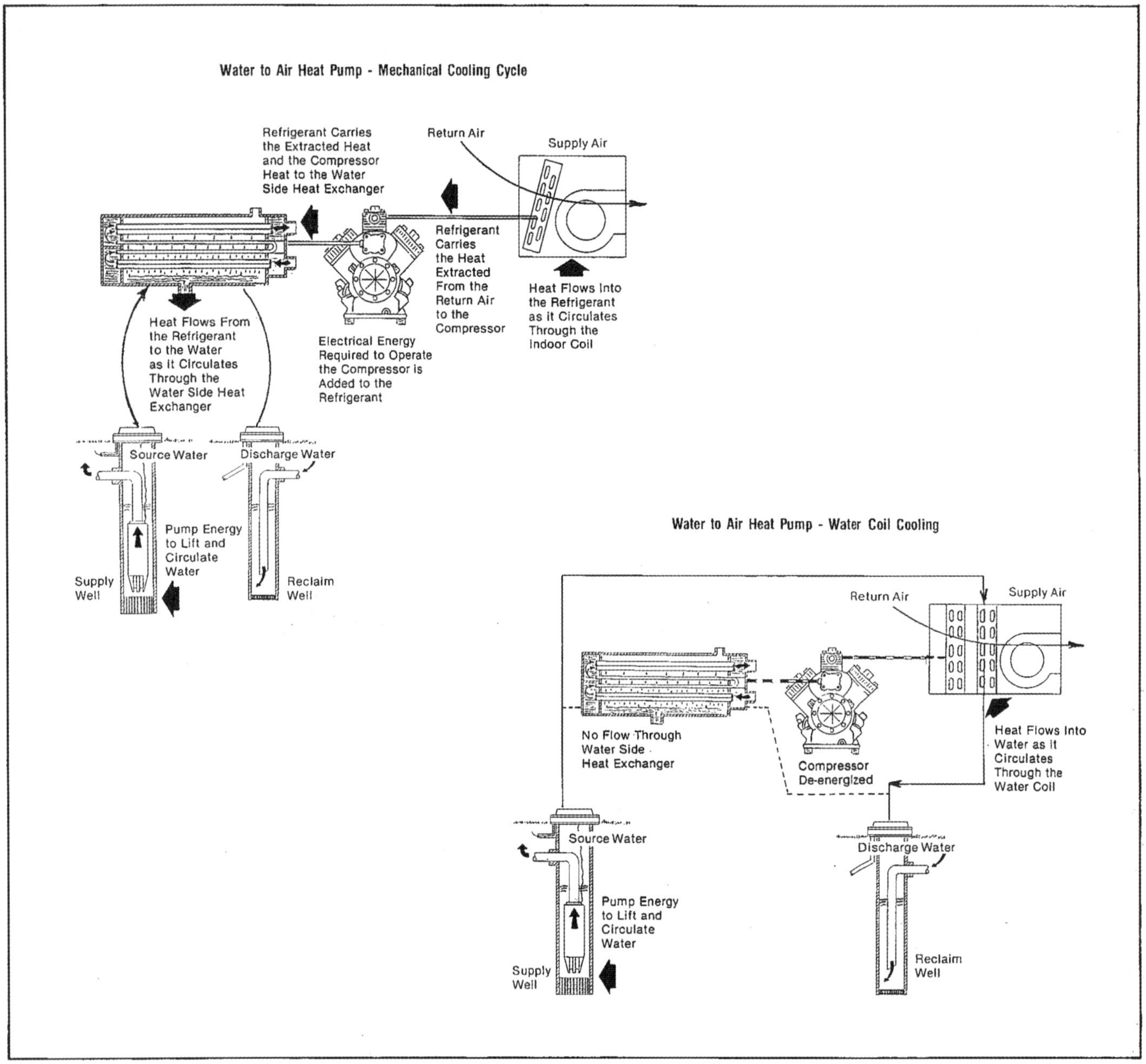

Figure 2.4 Cooling Cycle Energy Flow

Figure 2.4 indicates how the equipment operates in the cooling mode.

In general, the ground water equipment is able to operate at higher heating and cooling efficiencies than similar air to air equipment. However, when the equipment performance is evaluated, the energy consumed by the well pump must be included. Some manufacturers do not include this pumping energy in their catalog performance data. The overall efficiency (COP or EER) will be reduced when the pumping energy is included. A 10 percent to 20 percent (in some cases up to 50 percent) reduction in heating COP, and a 5 percent to 15 percent (up to 30 percent) reduction in cooling EER is possible.

$$\text{COP}^* = \frac{\text{Btuh Output}}{(\text{Heat Pump Input Watts} + \text{Well Pump Watts}) \times 3.4\ \text{Btuh/Watt}}$$

$$\text{EER}^* = \frac{\text{Btuh Output}}{(\text{Heat Pump Input Watts} + \text{Well Pump Watts})}$$

*Published COP and EER values are established at a test stand under closely controlled conditions. Performance in the field will vary with installation and operating conditions.

Note that the practice of comparing air source heating COP with water source heating COP can be misleading because, during heating, the water source equipment has a fairly constant COP, while the air source heat pump COP varies with outdoor temperatures (refer to Figure 1.4). A valid comparison of air and water source performance should consider the incremental efficiencies and number of operating hours at the various outdoor temperatures. Figure 2.5 compares the heating performance characteristics of the two types of water source equipment with the performance characteristics of air source equipment.

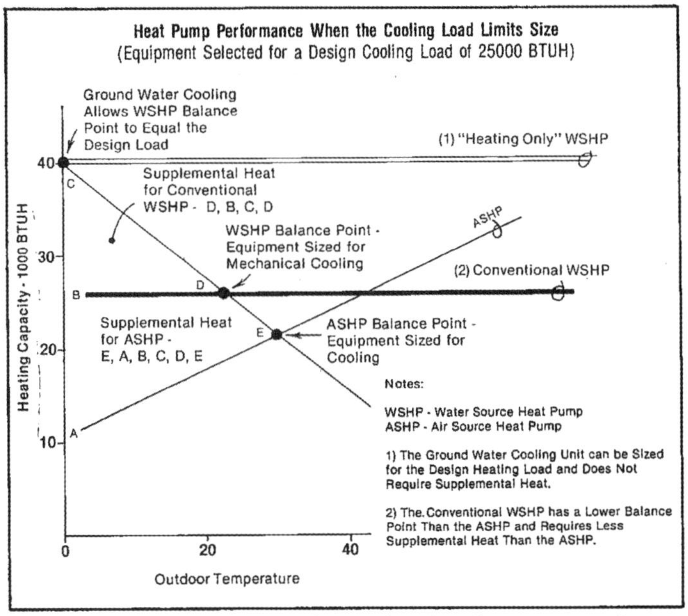

Figure 2.5 Comparison of Heating Performance of ASHP and WSHP Equipment

One important advantage of the water source equipment is that the defrost cycle is not required and, therefore, the efficiency of the equipment is not reduced. Compressor life should also be enhanced by the absence of the defrost cycle.

In the cooling mode, a water source heat pump can often provide mechanical cooling at a higher EER than an air source unit. This is possible because the condensing temperature of the water source equipment is considerably lower than that of the air source unit. However, some of this advantage is offset by energy consumed by the well pump. It is difficult to make specific statements about the differences in cooling efficiencies because the EER and SEER ratings for both air source and water source equipment vary considerably with make and model. It is reasonable to say that a water source heat pump that is designed for water coil cooling will operate at a very high cooling EER because the compressor is de-energized and only the water pump and fan power are required. In this case, a cooling EER of 15 to 25 or higher is possible depending on water temperature, pumping power and fan energy requirements.

Figure 2.6 compares the cooling performance characteristics of the two types of water source equipment with the performance characteristics of air source equipment.

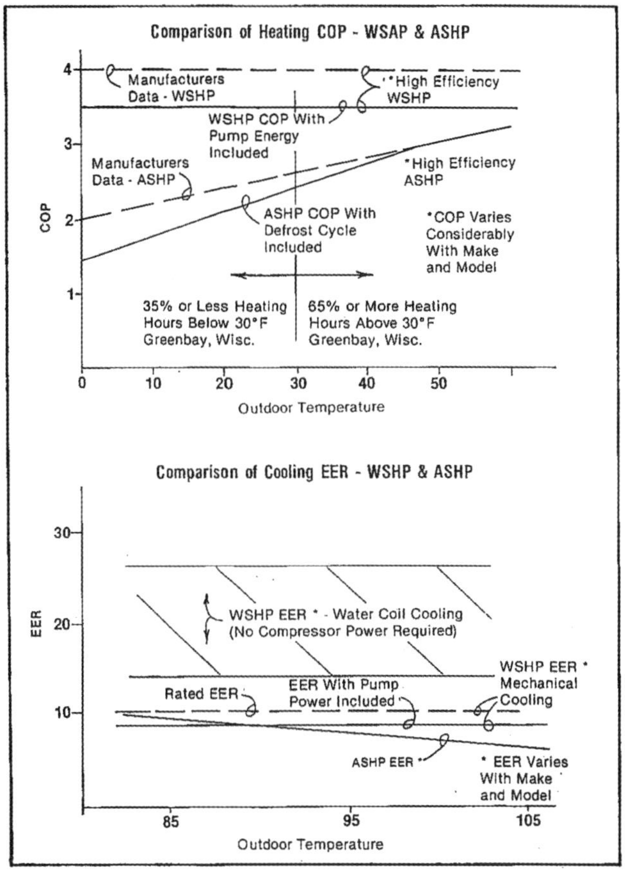

Figure 2.6 Comparison of Cooling Performance WSHP and ASHP Equipment

The disadvantages of the ground water heat pump are mostly associated with developing the water source and the problems associated with water disposal. Since two to four GPM of ground water are normally required to provide 12,000 Btuh of heat, a residential application can require a water flow rate of four to ten GPM and the flow required for commercial work will be considerably more. The production and disposal of this much water becomes a primary design consideration.

Wells are the most common source of ground water. But a lake, pond, or stream can sometimes be used for a water source or for water disposal. Normally, two wells are required. One well serves as a source and the second is used for disposal. Care must be exercised to avoid thermal coupling and both wells must be carefully designed so their supply and diffusion capacities can be established and maintained for many years. (Return wells must be deeper and as large or, preferably larger, in diameter than the supply well and must have about twice the screen length of the supply well.

Return wells should be located at least 50 to 100 feet from the supply well.)

Another option is to install a two well reversible system. (This requires installation of a pump in each well.) The advantages of the reversible system are: 1) The wells are periodically redeveloped as the flow is reversed. 2) The wells may be spaced closer together. 3) There are some thermal benefits associated with reversing the source and the sink when switching from heating to cooling.

The concept of using a single well as a source and a sink has been proposed for situations where the aquifer and the geological formation does not produce enough water volume for a conventional supply well (Figure 2.7). This design is similar to the vertical earth coil system described in Section 2.10 except it is not a closed system. The single well design recycles the source and sink water in the vicinity not a closed system. The single well design recycles the source and sink water in the vicinity of the bore hole so that it can exchange heat with the surrounding geological structure. This arrangement can only be used in certain types of (consolidated) geological formations. Consult the National Water Well Association and the equipment manufacturer for detailed application information before attempting this type of installation.

Ground water systems also require a well pump, pressure tank, special valves and controls and other water side hardware. Flow control valves or orfices and automatic shut off valves should be included to avoid wasting water and energy. (The well driller is usually responsible for the well and the water system hardware.) Figure 2.8 illustrates the water side installation.

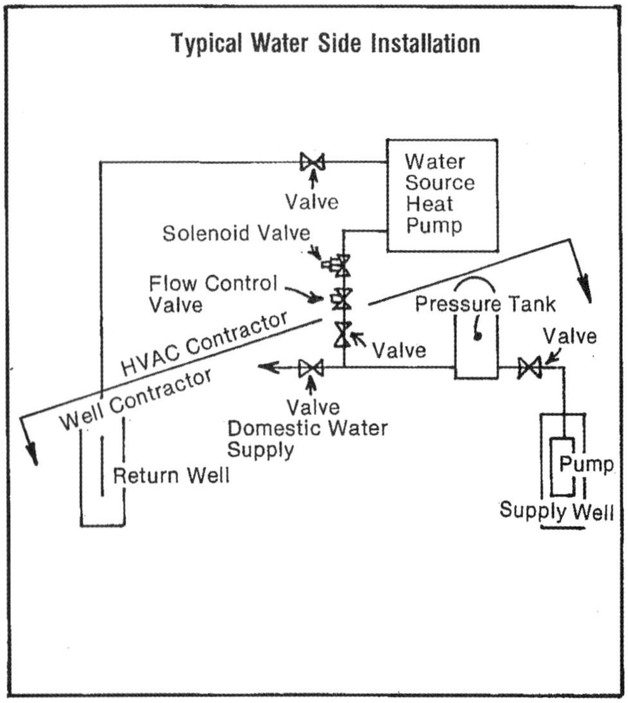

Figure 2.8

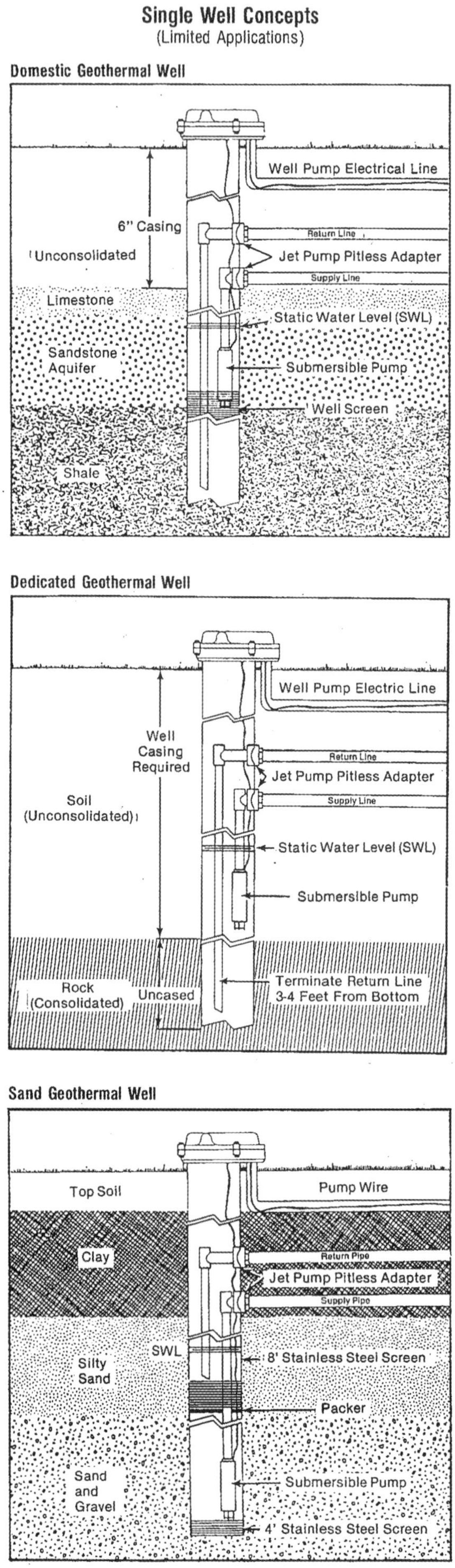

Figure 2.7

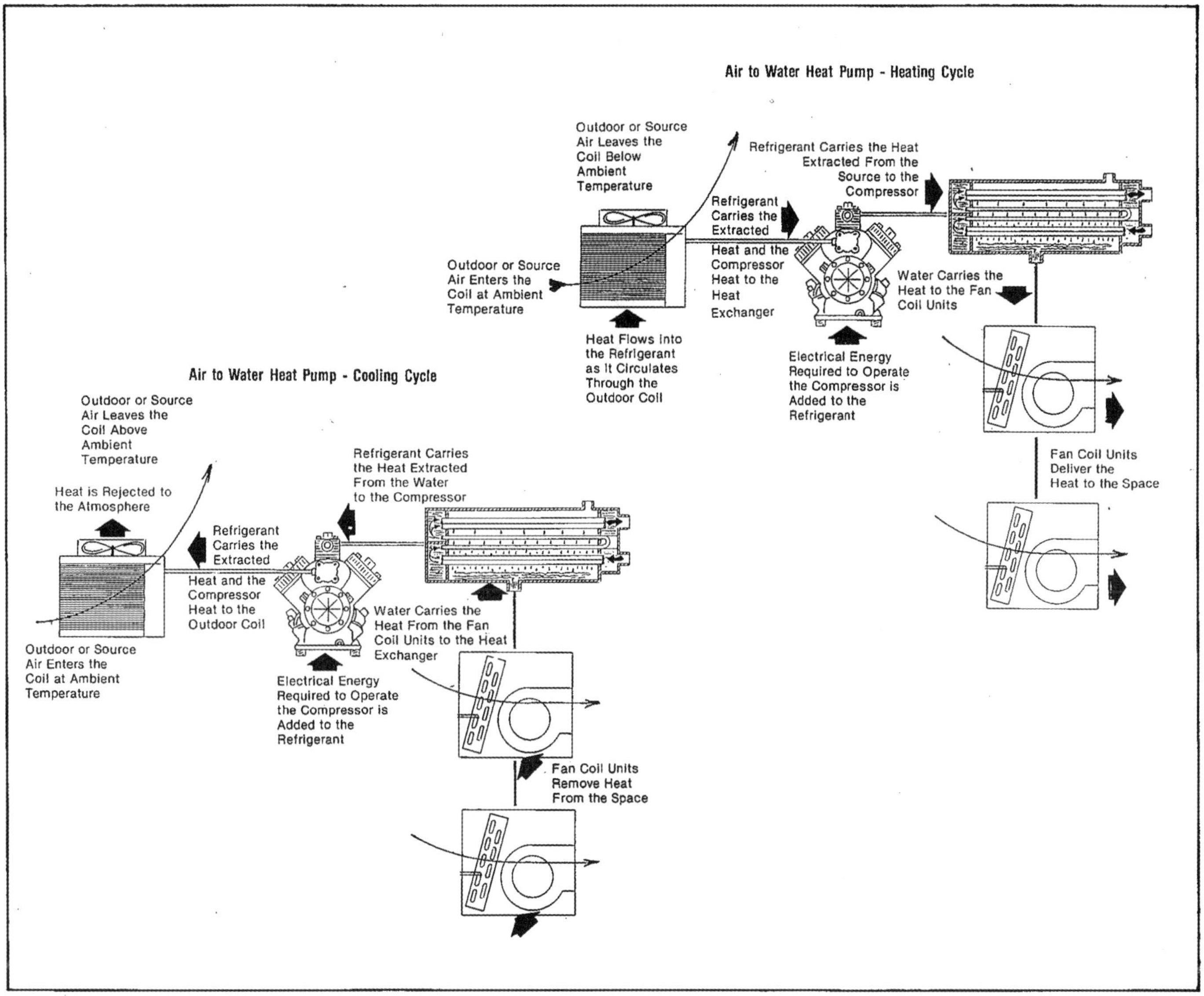

Figure 2.9 Air to Water Energy Flow

2.4 Air to Water Heat Pumps

These devices extract heat from the source air stream and deliver heat to water that is circulated through hydronic space heating system. The source air may be provided by the atmosphere or by a building or process exhaust system. Figure 2.9 illustrates the energy flow when the equipment operates in the heating or cooling modes.

This equipment is not as common as air to air or water to air equipment, but it is available in sizes that are suitable for single or multi-family residential and commercial applications. Because the occupied space is ultimately heated or cooled by hydronic devices, the air to water heat pump lends itself to designs where zone control is required. This is accomplished by allowing the zone thermostats to operate the various hydronic devices while the heat pump maintains a constant water supply temperature. The air to water heat pump can also be used as an "add on" heat pump to supplement an existing residential or commercial hydronic system. And, like any air source heat pump, this equipment is well suited to reclaim heat from building exhaust systems. The reclaimed heat can be used to supplement potable or process hot water systems or to provide space heating.

When fan coil units are used as terminal devices, they will provide adequate heating performance at relatively low supply water temperatures. Fan coil units also provide a means of introducing outdoor air and supplemental heat at each zone. In this case, the centrally located heat pump can be sized to take advantage of any diversity in the zone heating or cooling loads. Zone fan coil units selected to meet the greater of the heating or cooling requirement will insure adequate zone heating or cooling capacity. In this application, large duct systems are replaced by smaller water piping systems.

NOTE: The use of hot water radiation (convectors) as a terminal device is limited because convectors require temperatures that are higher than the 90°F to 100°F supplied by an air to water (R-22) heat pump. However, heat pumps that use other refrigerants, or are equipped with multistaged compressors, can produce water temperatures that are suitable for use with convection devices.

[The performance of the air to water equipment can be greatly enhanced by using compound (stated) compressors. Compound compression increases the amount of heat that can be extracted from a relatively low temperature source. When compared to a single stage unit of the same capacity, the compound equipment will extract heat at a lower power draw and operate at a higher COP. When compound compression is used, the supplemental heat requirement may be reduced or even eliminated. During cooling, when compound compression is not required, the compressors can be valved to operate as two single stage units in parallel. One disadvantage is that the multi-stage equipment is more expensive than the single stage equipment.]

Because air to water systems ultimately provide hot or cold water for heating or cooling, the system efficiency can readily be increased by including a provision for thermal storage. With such storage capability available, the heat pump can produce hot or cold water when outdoor conditions are more favorable and store it for later use. However, economic considerations may limit the number of applications where this approach is feasible.

The disadvantages associated with air to water equipment are that the COP is reduced as the outdoor temperature falls. Therefore, a defrost cycle is required when the outdoor temperature drops below 45°F. Also note that air to water equipment is subject to the same balance point limitations and supplemental heat requirements that apply to the more common air to air equipment. Air to water equipment will generally have a somewhat lower COP than similar air to air equipment because the heat must first be transferred to the hydronic system before it is delivered to the occupied space.

Since this equipment lends itself to specialized design applications, no general statements can be made in regard to first costs and operating costs. Such discussions require a comparison with alternate systems that satisfy the same design parameters. Application of this system requires a considerable amount of engineering to verify technical and economic feasibility and to optimize the final design configuration.

2.5 Water to Water Heat Pumps

This equipment extracts heat from the source water and deliver the heat to water that is circulated through hydronic devices. The source water may be provided by a well, lake or by a fluid discharge from a process. Figure 2.10 illustrates energy flow when the equipment operates in the heating or cooling modes.

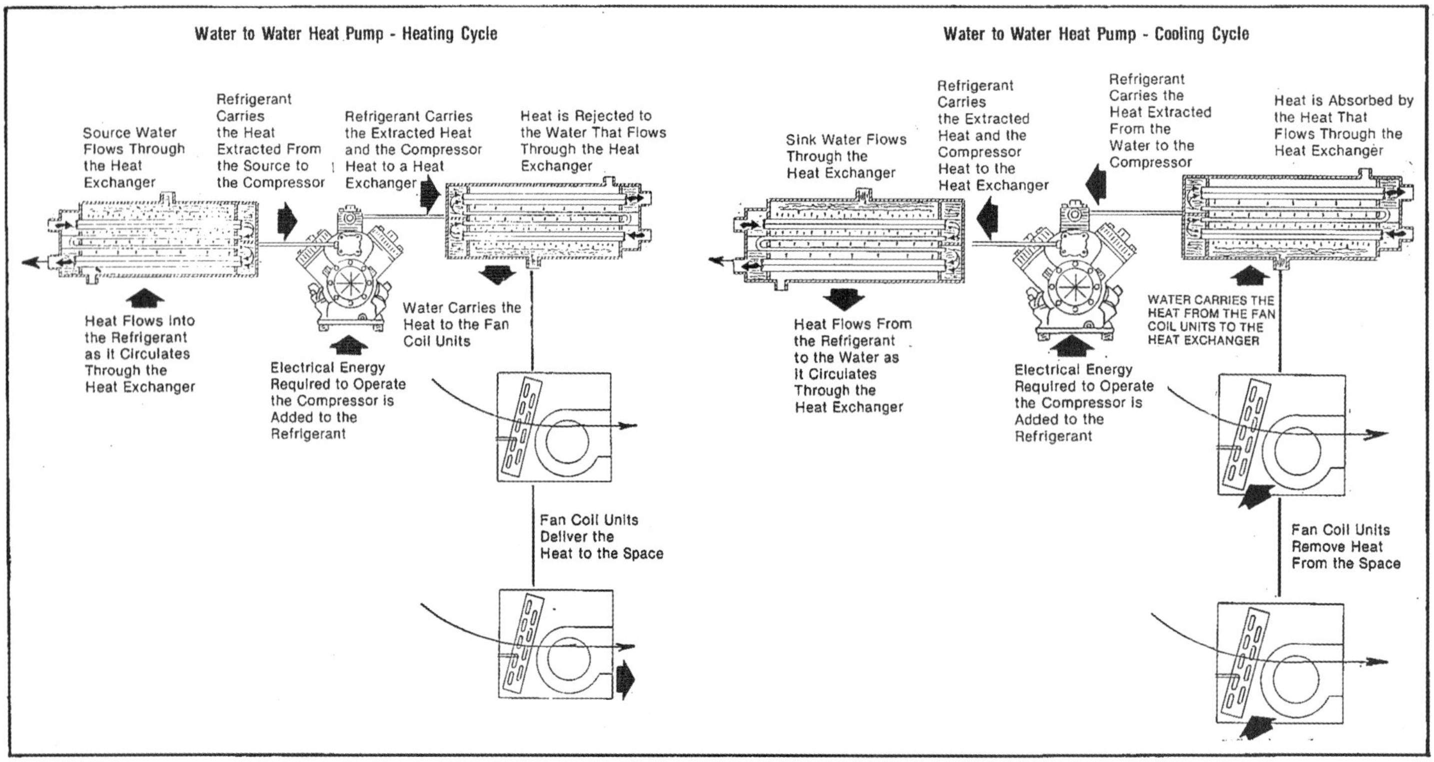

Figure 2.10 Water to Water Energy Flow

Water to water systems are not as common as air source or water to air equipment, but this equipment is available in sizes suitable for commercial, single and multi-family residential applications. This equipment also finds application in situations where a water to water reclaim system is desired.

Water to water heat pump systems can be zoned by equipping each zone with a fan coil unit which is capable of supplying outdoor air and supplemental heat. In this application, the centrally located heat pump can be sized to take advantage of any diversity in the zone heating or cooling loads. Large duct systems are eliminated, but a water piping system is required. Fan coil units selected to meet the greater of the heating or cooling requirement will insure adequate zone heating or cooling capacity.

The use of hot water radiation (convectors) as an hydronic terminal device is limited by the relatively low supply water temperatures that are produced by this equipment. Convectors are usually designed for operation at water temperatures that are higher than the 90°F to 110°F supplied by the water source heat pump. At additional cost, compound compressors can be used to elevate the supply water temperature and to increase the overall efficiency of the system.

This equipment, like any water source heat pump, is well suited to reclaim heat from waste water or fluids discharged from a manufacturing process. The reclaimed heat can be used to supplement potable or process water systems or for space heating.

This equipment has the ability to operate at a relatively high COP or EER and a defrost cycle is not required. The performance of the water to water equipment will be somewhat less efficient than similar water to air equipment because of the additional heat transfer process associated with the hydronic system.

The disadvantages of the water to water heat pump are related to the source of water supply. Installation costs and maintenance costs are increased, environmental factors are a consideration and pumping power will reduce the efficiency of the system. Refer to Section 2.3 for a more complete discussion of these factors. However, if this equipment is used to reclaim heat from a discharge of clean non-corrosive fluid, many of these disadvantages may be eliminated.

Since this equipment lends itself to a limited number of design applications, no general statements can be made in regard to first costs and operating costs. Such discussions require a comparison with alternate systems that satisfy the same design parameters. Application of this system requires a considerable amount of engineering to verify technical and economic feasibility and to optimize the final design configuration.

2.6 Hybrid Heat Pump Systems

Hybrid heat pumps are similar to ordinary heat pumps except that the hybrid heat pump is supplemented by a fossil fuel furnace or boiler. When the outdoor temperatures are moderate, the building heat requirements can be entirely satisfied by the heat pump. When the outdoor temperatures are below the economic balance point, the heat pump is de-energized and the fossil fuel heating system is used to heat the building. These systems are also referred to by such terms as "add on", "dual fuel", "optimizer and others.

The hybrid heat pump equipment offers the following advantages:

1) Full advantage can be taken of the high efficiency of the heat pump when the outdoor temperatures are moderate.

2) The inefficiency associated with cycling a fossil fuel furnace or boiler under low load conditions is eliminated.

3) When outdoor temperatures are low, the furnace or boiler supplies heat at close to its peak efficiency.

4) If the fossil fuel is relatively expensive (compared to electricity) the hybrid system will have a lower operating cost than a fossil fuel heating system.

5) If electricity is relatively more expensive than the fossil fuel, the hybrid system will operate at a lower cost than the all electric heat pump.

6) The heat pump can be retro-fitted to an existing electric or fossil fuel furnace and a more efficient system can be provided without disposing of equipment that is in good operating order.

7) The elimination of supplemental electrical resistance heating coils will reduce the peak demand on the electrical service system.

The hybrid system also has some limitations:

1) Lower operating cost is the most attractive feature of this system, but a careful analysis must be made to determine if it is the most economical option. Under some conditions, fossil fuel heating, electric cooling or all electric heat pump systems can operate at lower cost. An analysis which considers equipment efficiencies, fuel cost, weather data, and the characteristics of the building envelope and internal loads must be made to identify the most economical option.

2) Some gas utilities add surcharge to the monthly billing to compensate for providing full service at a reduced level of consumption.

3) When a heat pump system is added to an existing warm air system, care must be taken to insure that the existing fan, duct system and air terminals are compatible with the heat pump air side requirements. (This is discussed in more detail in the last paragraph of Section 4.3 and in Section 5.13.)

4) When fossil fuel is relatively expensive, the annual heating costs are minimized if both heat pump and the furnace can operate simultaneously. However, this is not a generally accepted practice because most manufacturers and many building codes require installation of the Indoor DX Coil down stream (in the supply duct) of a fossil fuel furnace. This location eliminates any possibility of operating the heat pump while the furnace is heating since this would result in high pressures in the refrigerant system. Refer to Section 3.10 for information on control strategy and Section 5.13 for information on installation procedures.

   Note that when codes are favorable and if simultaneous operation does not result in excessive heating coil operating temperatures, the heat Pump DX Coil may be located upstream of an electric furnace. In this position, the heat pump will be able to operate when the furnace is delivering heat to the space.

Hybrid systems have a thermal balance point and an economic balance point. (The thermal balance point has already been discussed in Section 1.6) The economic balance point occures at the outdoor temperature where the cost of heating energy supplied by the heat pump is equal to the cost of heating energy supplied by the fossil fuel. Hybrid system operating costs are minimized by switching to fossil fuel heat when the outdoor temperature is below the economic balance point.

The economic balance point depends on the heat pump COP, the furnace or boiler efficiency at various outdoor temperatures, the fossil fuel costs and the cost of electrical energy. The economic balance point can be estimated by calculating the heat pump COP that provides heat at the same cost as the fossil fuel furnace and using this COP with the manufacturers data to find the outdoor temperature at which this COP occurs. This COP can be estimated by using the following equations:

For Natural Gas

$$\text{Break Even COP} = \frac{1{,}000{,}000 \text{ Btu}}{3413 \text{ Btu/KW}} \times \frac{\$/\text{KWH}}{\$/1000 \text{ Cu. Ft.}} \times 0.9 \times \text{S.S. Furnace Eff.}$$

For Oil

$$\text{Break Even COP} = \frac{138{,}000 \text{ Btu}}{3413 \text{ Btu/KW}} \times \frac{\$/\text{KWH}}{\$/\text{Gal}} \times 0.9 \times \text{S.S. Furnace Eff.}$$

For Propane

$$\text{Break Even COP} = \frac{92{,}000 \text{ BTU}}{3413 \text{ BTU/KW}} \times \frac{\$/\text{KWH}}{\$/\text{Gal}} \times 0.9 \times \text{S.S. Furnace Eff.}$$

Where:

$ .... Dollars
S.S. .... Steady State
KWH .... Kilowatt Hours
Cu. Ft. .... Cubic Feet
Gal .... Gallons
Eff .... Efficiency
0.9 is a factor to account for furnace cycling.

These equations are reasonably accurate for furnaces that are not severely oversized. If the heat pump "break-even" COP is related to the manufacturers integrated capacity data, the procedure will account for the losses associated with the defrost cycle. (When possible, an oversized furnace should be modified so that its capacity is as close as possible to the design heating load.) These equations can also be used to make a quick check on the economic feasibility of using the add on system. Consider the following situations:

Case 1

Gas Furnace Efficiency = 80%
Gas Cost = $4.50/MCF

Electrical Energy Cost = $0.07 Per KWH

$$\text{COP} = \frac{1{,}000{,}000}{3413} \times \frac{.07}{4.50} \times 0.9 \times 0.8 = 3.28$$

Such a high COP could only be realized at an outdoor temperature above 60°F. An economic balance point of 60°F indicates that the add on heat pump will not produce a significant savings.

Case 2

Oil Furnace Efficiency = 70%
Oil Cost = $1.25/Gal

Electrical Energy Cost = $ 0.05 per KWH

$$COP = \frac{138{,}000}{3413} \times \frac{.05}{1.25} \times 0.9 \times 0.70 = 0.96$$

This COP is obtainable even when the outdoor temperature is below 0°F. An economic balance point in this range indicates that the add on unit will produce a significant savings.

Case 3

Gas Furnace Efficiency = 80%
Gas Cost = $5.00/MCF

Electrical Energy Cost = $ 0.05 per KWH

$$COP = \frac{1{,}000{,}000}{3413} \times \frac{.05}{5.00} \times 0.9 \times 0.8 = 2.1$$

This COP will occur when the outdoor temperature is near 30°F. An economic balance point in this range indicates that the hybrid system will effectively reduce operating costs (in comparison to an all electric heat pump system) if that equipment is located in an area that has a high number of heating degree days.

In summary, the hybrid system is favored when fossil fuel is relatively inexpensive in comparison to electricity and is more effective in colder climates. The all electric heat pump is favored if electricity is relatively inexpensive or if the climate is mild.

Other hybrid applications include air to water, or water to water heat pumps which are combined with a fossil fuel boiler to produce a hybrid hydronic system. This "heating only" application uses a refrigerant to water heat exchanger to transfer the heat from the heat pump to the hydronic system. This system is subject to the economic balance point considerations disucssed above and to other consideratons outlined in Sections 2.4 and 2.5. Figure 2.11 illustrates the air to water hybrid hydronic system.

Hybrid systems may not always be the result of an "add-on" application. A new type of hybrid system has recently been introduced in a configuration that includes a gas burner and a heat pump. The equipment arrangement allows the heat pump and the gas burner to operate simultaneously. This is accomplished by locating a gas burner beneath the "A" coil in the outdoor unit. When the outdoor temperature is above 40°F, the unit operates strictly as a heat pump. Below 40°F, the gas burner is energized and the heat pump continues to operate. In this mode, the

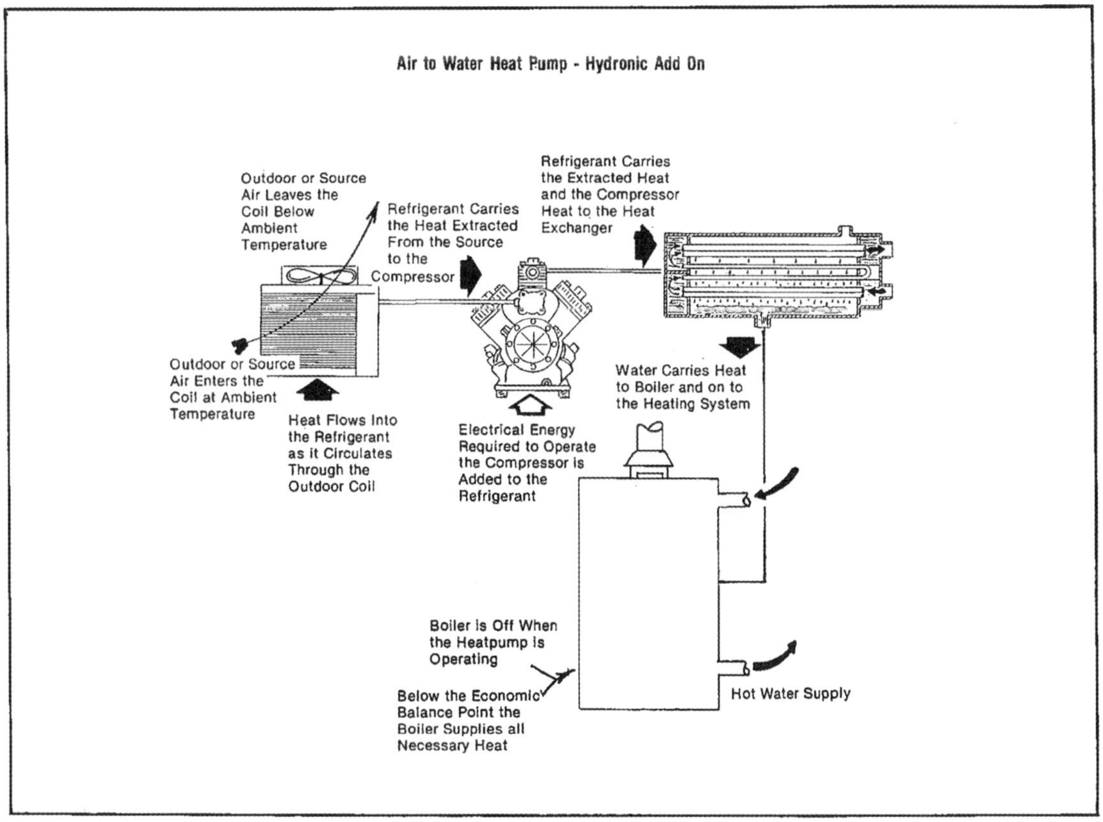

Figure 2.11 - Air/Water Hydronic System

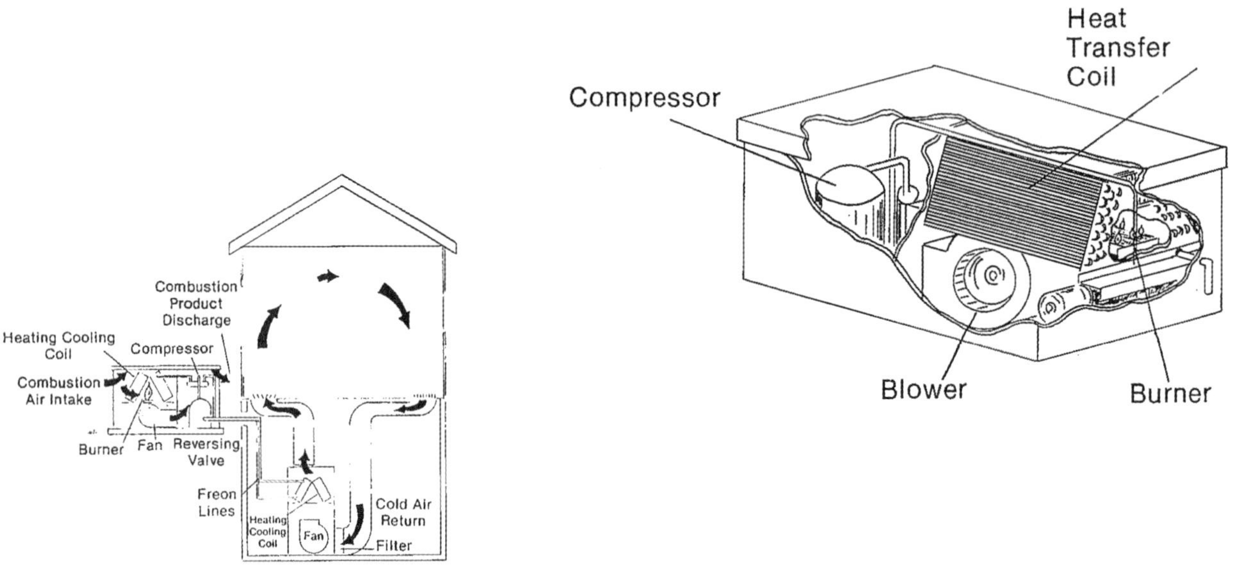

Figure 2.12 Hybrid Unit

heat pump captures the heat from the burner and delivers it to the indoor unit. Refer to Figure 2.12 for an illustration of how this equipment operates.

This design has the following advantages:

A) When the outdoor temperature is above 40°F, the heat pump supplies all the necessary heat at a relatively high efficiency.

B) When the outdoor temperature is below 40°F, the heat pump continues to operate. In this case, the burner is firing and the products of combustion are allowed to condense on the outdoor "A" coil. The flue gases exit at temperatures near 60°F and the overall efficiency is comparable to that of a high efficiency furnace.

C) During cold weather, the heat pump operates at a 60°F coil temperature and defrost is not required. Since the burner is outdoors, no flue is required.

2.7 Reclaim Heat Pump Systems

Heat pumps are among the most efficient and versatile devices available for reclaiming heat that would otherwise be wasted. Air source heat pumps can be used to reclaim both sensible and latent heat. Water source heat pumps can reclaim heat from waste water or other liquid discharges. They can be used to transfer heat from zone to zone within the same building.

One distinct advantage that all reclaim heat pump systems enjoy is that they are able to extract heat from sources which may be too cold for use with other types (heat wheels, refrigerant coils, air to air exchangers) of reclaim equipment.

Air to air heat pumps can be used to temper outdoor (ventilation) air with heat reclaimed from exhaust air. This application has the following advantages:

A) Both sensible and latent heat can be recovered.

B) Cross contamination of the air streams is eliminated.

C) Side by side installation of the intake and exhaust air ducts is not necessary. However, the distance between the refrigerant coils which must be installed in the intake and exhaust ducts will be limited by the pressure drop which occurs in the refrigerant piping.

Air to water heat pumps can recover sensible and latent heat from exhaust air and deliver that heat to a potable or hydronic water heating system. These units can also be used in conjunction with a closed water piping loop to reclaim and transfer heat from zone to zone within a building. Refer to Section 2.8 for more information on water loop systems.

Water to air and water to water heat pumps can be used to reclaim heat from waste water or fluids expelled from industrial processes and deliver that heat to a process or comfort heating system.

Reclaim heat pump applications are only limited by the designer's imagination. However, every application requires a considerable amount of engineering and analysis to insure that the design is technically and economically feasible.

Table 2.0 Applications for Heat Reclamation Heat Pumps

Energy Source	Temper Ventilation Air	Preheat Domestic Hot Water	Space Heating	Terminal Reheat	Temper Makeup Air	Preheat Combustion Air	Heavy Oil Heating	Internal To External Zone Heat Transfer
Exhaust Air	A/A	A/W	A/A A/W	A/W	A/A	A/A	A/W	
Process Exhaust & Computer Rooms	A/A	A/W	A/A A/W	A/W	A/A	A/A	A/W	
Hot Condensate	W/A	W/W	W/A W/W	W/W	W/A	W/A	W/W	
Refrigerant Hot Gas	A/A	A/W	A/A A/W	A/W	A/A	A/A	A/W	
Hot Condenser Water	W/A	W/W	W/A W/W	W/W	W/A	W/A	W/W	
Hot Water Drains	W/A	W/W	W/A W/W	W/W	W/A	W/A	W/W	
Process Waste Fluids	W/A	W/W	W/A W/W	W/W	W/A	W/A	W/W	
Air Compressor Cooling System	W/A	W/W	W/W	W/W	W/A	W/A	W/W	
Lights and Interal Loads								W/A Loop

A/A - Air to Air   A/W - Air to Water
W/A - Water to Air   W/W - Water to Water

A) Reclaim systems should only be considered when the reclaimable heat is both available and usable.

B) It is necessary to verify that a system energy balance exists at every operating condition.

C) Provisions for rejecting excess heat or for providing supplemental heat are usually required.

D) Provisions for thermal storage may be required and the effects of thermal storage on the system efficiency and the project economics must be evaluated.

E) The economics of the proposed heat pump reclaim system should be compared to alternate systems to verify that it offers the best return on investment.

Table 2.0 summarizes some of the more common reclaim heat pump applications.

2.8 Water Loop Reclaim Systems

Water loop reclaim systems employ water source (water to air) heat pumps to heat and cool rooms, or zones, and to transfer heat from area to area within the building. This special application of the water to air heat pump has become fairly common and involves considerations which were not covered in Section 2.2. Figure 2.13 illustrates the operating relationship between the basic system components which consist of heat pumps, heat source, and heat rejection device.

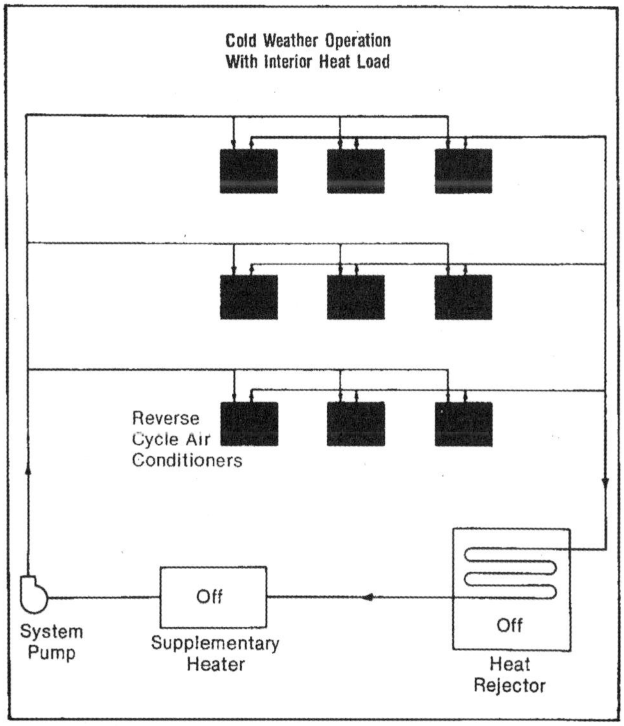

Figure 2.13 Water Loop System Operating Modes

These systems are usually designed to maintain the water loop temperature between 60°F and 90°F. When the loop temperature falls below 60°F, the supplementary heater is energized. When the loop temperature rises above 90°F, the heat rejector operates. Individual units add or remove heat from their respective zones as they reclaim heat from or reject heat to the water loop. When a unit operates in the cooling mode, the heat extracted from the zone plus the heat of compression is transferred to the water loop. When the unit operates in the heating mode, the unit extracts heat from the water loop, adds the heat of compression and delivers this heat to the zone. Note that only the heat reclaimed from a zone that requires cooling can be considered to be "free", and that the heat of compression or supplementary heat must ultimately be purchased. Figure 2.14 illustrates how the individual units operate.

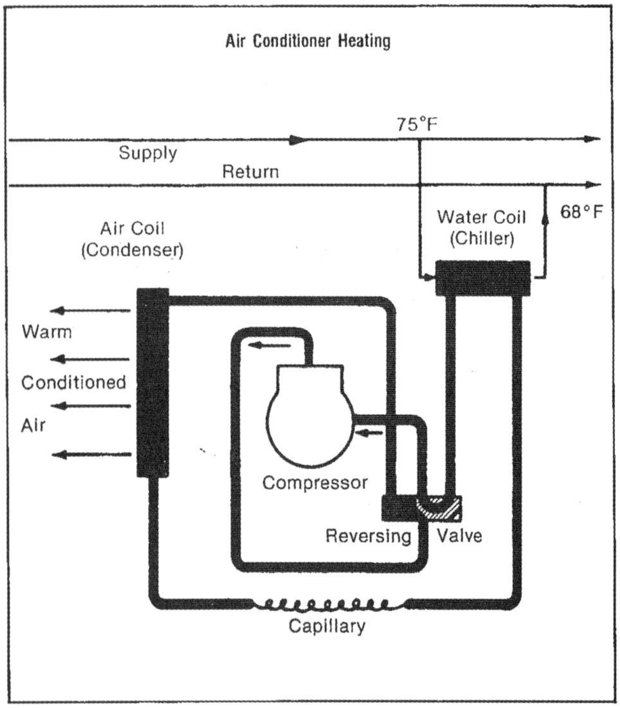

Figure 2.14 Water Loop Unit Operation

The heat pump units are available in many configurations. Small units required for rooms or small zones are available in a console, horizontal, or vertical arrangement. These units do not include a provision for introducing outdoor air into the space. Unitary equipment and unit ventilators are available which do include a provision for introducing outdoor air. Rooftop units which serve large zones are available and can be obtained with outdoor air and exhaust air dampers.

One unique feature of this system is that it can share the same piping used for the building sprinkler system. The system must be designed in accordance with the NFPA Standard 13, Chapter 5. This integrated system lowers the overall construction costs, reduces the accumulation of sediment in the fire protection system and increases the thermal storage capacity of the piping loop.

This system achieves its maximum efficiency when simultaneous heating and cooling is required, and when one third of the units operate in the cooling mode while the remaining two thirds provide heat. When the building is occupied, this ratio may be weighted toward cooling. In this case, the addition of thermal storage tank will usually improve the overall efficiency of the water loop system. (Thermal storage is necessary whenever the need for the reclaimed heat does not occur when the heat is available.) Buildings which have cooling loads in the perimeter zones during cold weather have a unfavorable ratio of simultaneous heating and cooling and are not good candidates for the water loop system.

A provision for an outdoor air economizer cycle can also improve the overall system efficiency. Economizer cooling should be employed when the outdoor temperatures are suitable and all the units are in the cooling mode. However, operating the economizer when the building requires simultaneous heating and cooling will reduce the efficiency of the reclaim system. In this case, the heat required by the perimeter zones must come from the heat reclaimed from interior zones and this heat must not be expelled through the economizer.

This system should be considered when a building has a design cooling load of fifty tons or more, a sizable internal load and a simultaneous heating and cooling requirement for a good part of the year. Like any energy conservation system, it requires a considerable amount of engineering to correctly size the components, design the control system, and to compare this system against other alternatives. First costs may be 10 to 20 percent higher than more conventional systems and an economic analysis is required to evaluate the return on investment.

2.9 Zoned Heat Pump Systems

Zoned systems can adjust to variations in individual room or zone heating and cooling loads. Furthermore, the zoned system has the ability to eliminate cooling or reduce heating in unoccupied spaces. Air to air heat pumps can be zoned by using multiple indoor coils in conjunction with a single outdoor condensing unit or by adding volume control dampers to a standard single zone air distribution system. Both types of systems are currently available. By taking advantage of load diversity within the building, these systems have the potential to maintain comfort while reducing energy usage.

Figure 2.15 illustrates a proprietary heat pump system that is zoned by using multiple indoor coils. The indoor units can be operated in groups or individually to maintain the setpoint of the selected zones. Two to five indoor units can be used in conjunction with one outdoor unit to condition occupied spaces ranging from 500 to 1500 square feet. Since the indoor units will not operate simultaneously, the outdoor condensing unit can have a capacity that is less than the sum of the individual indoor unit capacities. Capacity control is required on the outdoor unit so that it can operate under low load conditions.

In some models, supplemental heat, in the form of resistance strip heat, is installed in the indoor units. No duct work is required. Refrigerant piping runs have verticle rise and horizontal length limitations which are specified by the manufacturer. The indoor units do not include a provision for introducing outdoor air into the occupied space.

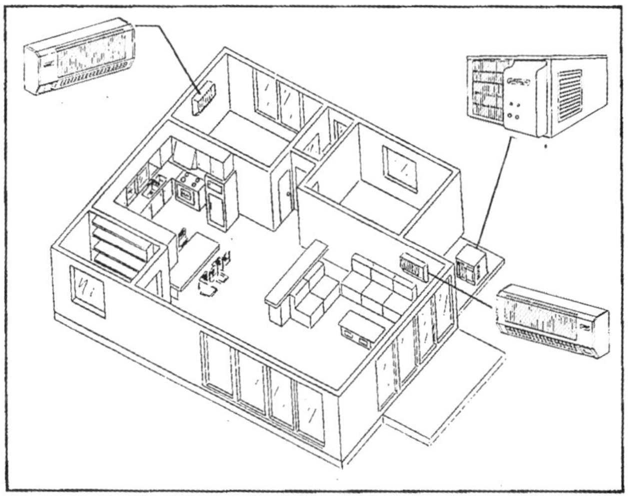

Figure 2.15 Zoned Refrigerant Side System

Figure 2.16 illustrates how a standard air to air heat pump system can be zoned by adding volume control dampers to the air distribution system. This arrangement essentially converts a single zone system into a variable volume system. In this design, the zone thermostats modulate zone dampers to maintain the zone setpoints. The air flow is throttled to those zones which are satisfied and zones which require heating or cooling receive maximum air flow. If all zone thermostats are satisfied, the control system de-energizes the compressor and opens all the zone dampers to re-establish air flow to all zones. Minimum air flow is maintained across the indoor coil at all points of operation by:

1) Sizing the air distribution system so that adequate flow can be maintained through any single zone.
2) Relieving excess supply air from the supply duct to the return system.

Outdoor air and supplemental heat can be provided at the central unit. Anti-freeze controls are required to protect the indoor coil if a low air flow condition occurs and capacity control, including hot gas, bypass is recommended. Protection against compressor short cycling is also recommended.

Commercial applications require consideration of all problems associated with any variable volume system. These include, but are not limited to, zone air movement and ventilation requirements, provisions for simultaneous heating and cooling, air terminal performance as supply CFM varies, DX Coil Control, and poor performance when equipment is oversized. Residential applications require careful design of the duct system to insure adequate air flow over the indoor coil at all points of operation.

2.10 Earth Coupled Heat Pump Systems

Earth coupled heat pump systems employ a closed water loop that acts as an intermediate heat exchanger which absorbs heat from the ground and delivers that heat to a water source heat pump. The water to air heat pump extracts heat from the water loop and delivers the heat to the occupied space. When properly installed, this system may be more efficient than an air source heat pump, but not as efficient as a ground water heat pump.

The closed water loop must be buried in contact with moist earth or immersed in a well. The water loop is filled one time with potable water and no ground water is circulated through the system. This closed, sealed system eliminates the problem of water side deposits forming on the system piping and heat exchanger. Another advantage is that the heat pump system is physically isolated from the aquifer. In addition, the closed water loop requires only a small circulating pump because the lifting head is eliminated and the pump only has to overcome pipe friction losses.

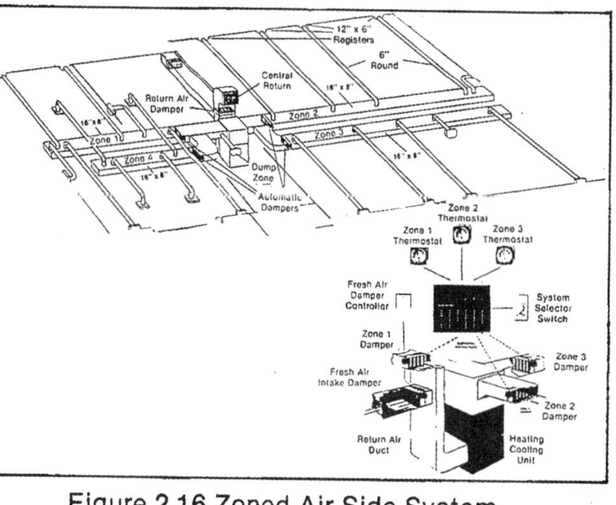

Figure 2.16 Zoned Air Side System

These systems must be designed to balance the heat removal and recovery capabilities of the loop with the design cooling and heating loads. In addition, allowances must be made for seasonal variations in the average loop temperature that can be as high as ± 20°F. (Antifreeze is required if the loop temperatures fall below 40°F.)

The two types of water loop systems which are available are illustrated by Figure 2.17.

The vertical earth coupled well is a closed water piping loop that is installed in an ordinary well hole. The depth of the well depends on the heat pump capacity and the geographic and geological conditions. The plastic pipe which forms the loop extends down into the aquafier, and the well hole is backfilled with sand or small gravel. The bore holes should be at least 4 3/4 inches in diameter and 15 feet apart. This type of system does not require a great amount of acreage for installation.

In the horizontal system, plastic pipe is buried at a depth of four to six feet. The pipe is serpentined to form multiple rows and, in some cases, a "soaker" line is used to wet the soil for better heat transfer. This system requires a considerable amount of acreage to accomodate the buried loop.

The installation costs of the earth coupled systems are relatively high and there is a shortage of design and application information. (Wide regional variations in the soil composition and water table levels limit the accuracy of the information that is available.) Another disadvantage is that a buried piping system cannot be easily inspected and repaired.

Vertical Earth Coupled Borehole and Piping

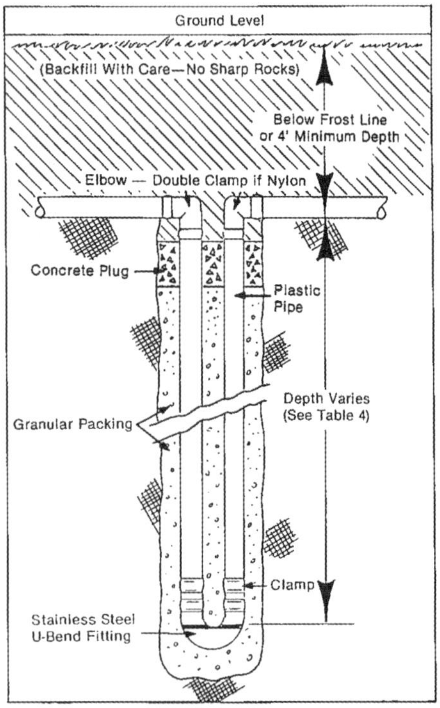

Horizontal Earth Coupled Pipe Separation in the Trench

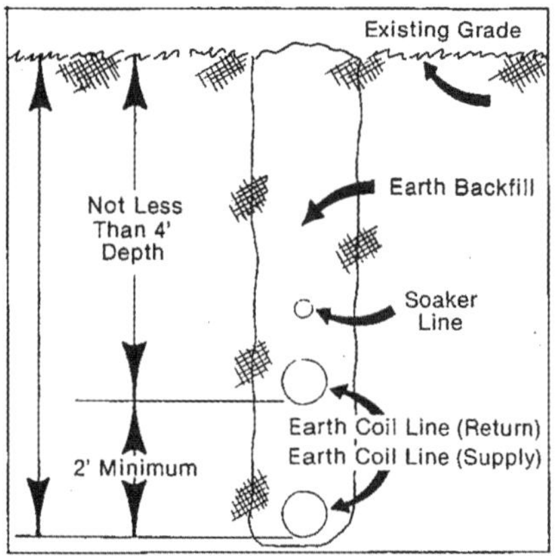

General Guidelines For Vertical Earth Coils

Avg. Well Water Temp.	Minimum Wetted Borehole Linear Ft/Ton*
80	190
75	180
70	170
65	160
60	150
55	170
50	190
45	220
40	260

*Based on 1-1/2" polybutylene or high density polyethylene

Avg. Well Water Temp.	Min. Depth (Ft)	Min. Pipe Length Linear Ft/Ton*
80	4	500
75	4	475
70	4	450
65	4	425
60	4	400
55	5	450
50	6	500
45	NA	NA
40	NA	NA

*Based on 2" polybutylene or high density polyethylene pipe.

Figure 2.17

2.11 Solar Assisted Heat Pumps

Solar assisted heat pump systems combine solar heating equipment with a water source or air source heat pump. These systems are capable of operating in many different modes. When the sun is intense, it may be possible for all heating to be provided by solar collectors. When the sun is less intense, solar collectors can be used to supplement the heat pump. The heat pump can also operate with no help from the solar system. Thermal storage is desirable because it provides a means to save excess heat captured by the solar panels or heat rejected from the cooling cycle for use at a later time.

In one design, the solar heated water is fed to a storage tank which provides a source of heat for the heat pump. If the storage water temperature should drop below 60°F, the auxiliary heat can be energized. During cooling, the system works in reverse, rejecting the heat of condensation to the storage tank until the tank temperature reaches its allowed maximum, then rejecting the heat to the sink.

Other designs place a "solar" hot water coil at the inlet side of the outdoor coil (where it can supplement the heat pump) or at the outlet side of the indoor coil (where it can provide heat when the heat pump is not operating), or in both the indoor and outdoor units to provide a combination of operating modes.

Figure 2.18 illustrates the designs described above. Many other configurations are also possible.

Solar assisted heat pumps require a considerable amount of design and engineering. They require sophisticated controls and are very expensive to install. The incremental gain in efficiency over conventional heating systems will usually not justify the higher installation costs of the solar system.

2.12 Heat Pumps Added to Wood or Coal Furnaces and Boilers

ACCA does not recommend adding a heat pump to a solid fuel furnace air distribution system. However, a hot water heating coil located down stream from the heat pump could conceivably be applied to a heat pump air distribution system. In this case, the supply air fan must be large enough to overcome the air side pressure drop through the coil. The heating coil can be controlled by using ordinary supplemental heating controls to energize a hot water circulating pump or to operate a valve.

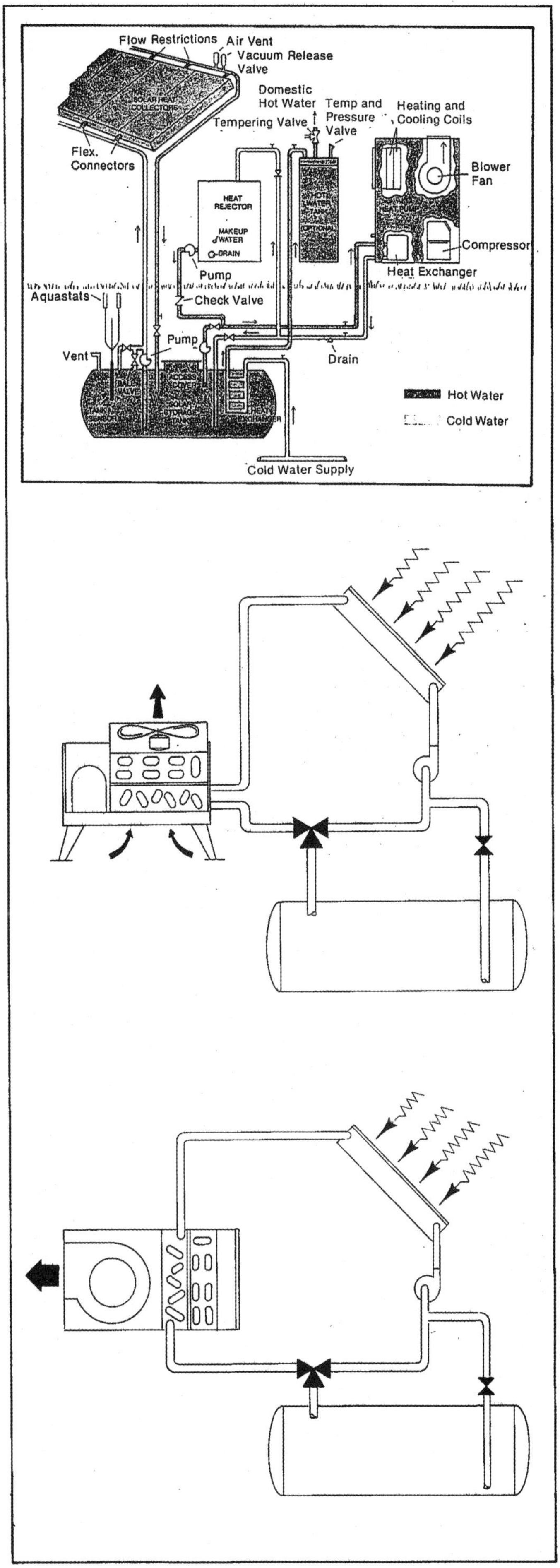

Figure 2.18 Solar Assisted Heat Pumps

2.13 Heat Pump Water Heater

The air source heat pump water heater (Figure 2.19) is a device that extracts heat from surrounding air and transfers that heat to the domestic water heating system. This device is a non-reversing, water heating device. But, cooling of surrounding air occurs as a by-product of the water heating operation.

There are two basic types of air source heat pump water heaters: the integral type and the add-on unit. The integral type can be purchased as one piece of equipment which replaces the conventional water heater. The add-on unit is designed to be connected to a new or existing water heater.

Since these devices extract heat from the surrounding air, they should be located in areas where the heat can be extracted without overcooling the space. In addition, they must be located in areas where ambient temperatures remain within the operating range of the equipment. They are particularly effective when waste heat is rejected into the surrounding space. Typical locations suitable for installing the heat pump water heater are:

Unconditioned spaces (temperature ranges 45°F to 95°F):

- Basement
- Crawl spaces
- Attached garages

Conditioned spaces:

- Basements
- Utility room
- Ventilated storage spaces

If the heat pump water heater is installed in a conditioned space, the heat that it extracts must be supplied by the space heating system, thereby increasing the space heating load. The net effect on the total heating energy costs depends on equipment efficiencies and the relative cost of fuels. However, during the cooling season, the heat extracted for water heating is essentially free and there is an added benefit of cooling and dehumidification in the space.

Air source heat pump water heaters should not be installed where ambient temperatures exceed 95°F or fall below 45°F. High ambients strain the compressor and low ambients result in coil frosting. (These units are not equipped with a defrost cycle.) An adaquate flow of air around the the unit is also required. Therefore, installation in small unvented rooms is not recommended.

Water source heat pump water heaters are also available. These devices can extract heat from ground water or reclaim heat from waste liquids.

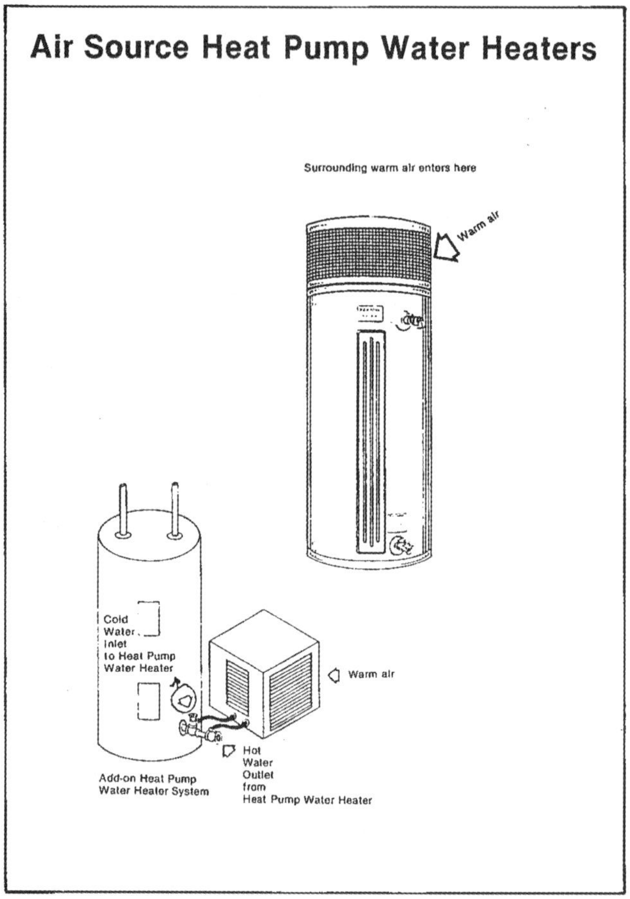

Figure 2.19 Air Source Heat Pump Water Heater

2.14 DeSuperheater Water Heaters

Heat pump de-superheaters are not heat pumps but accessories that are sometimes used in conjunction with air to air heat pump equipment. Desuperheater water heaters reclaim waste heat from the heat pump refrigeration cycle and use it to supplement the domestic water heater.

These units are very effective when an air to air heat pump is operating in the cooling cycle. In this case, the hot water produced by the unit is essentially free because the heat of condensation would normally be rejected to the sink. (They could also be used with cooling only equipment.) They can also produce savings when the heat pump is in the heating mode and operating above the balance point. No energy is saved below the balance point because all the heat pump capacity is required for space heating. The desuperheater has an added advantage in that it may produce an increase in the heat pump (cooling cycle) efficiency when the outdoor temperature, (condensing temperature) is high.

Desuperheaters have both commercial and residential applications but they are more effec-

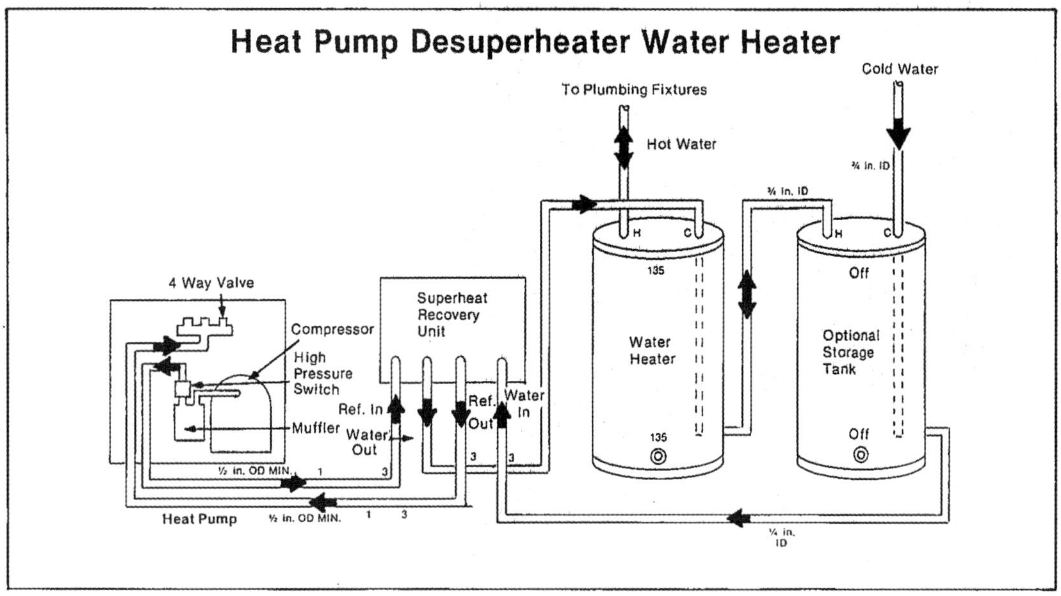

Figure 2.20 Desuperheater Water Heater

tive when warm climates or internal loads produce a large number of cooling hours and/or heating operating hours which fall above the balance point of the heat pump system.

Figure 2.20 illustrates the equipment arrangement.

The energy savings and cost effectiveness of the desuperheater water heater depend on the performance characteristics of the heat pump, the building heating and cooling requirements, the hot water demand, local weather patterns, control strategies, and thermal storage provisions.

Be sure to check for conformance with local codes and for the effect on manufacturers warranty agreement before installing this equipment.

Section 3

Principles of Heat Pump Operating Cycles

All types of heat pumps operate in accordance with the same principles which govern the basic refrigeration cycle. This is true whether the equipment is operating in the heating or the cooling mode. In every case, the basic components of a heat pump system will include a compressor, condenser, evaporator and refrigerant metering device. However, the heat pump does have the unique ability to switch the functions of the evaporator and the condenser as the equipment changes over from heating to cooling. Additional refrigerant side components such as a reversing valve, accumulator, check or metering valves are required to compensate for the differences in refrigerant flow rates and operating temperatures/pressures when the unit switches its mode of operation.

Differences between heat pump systems can be traced directly to the different types of heat sources and heat sinks. For example, air source heat pumps usually require defrost equipment, but defrost is not required on water source equipment. The use of DX coils, shell and tube heat exchangers, or intermediate heat exchangers will depend on the interface requirements between the source or sink and the heat pump refrigeration cycle. Of course, auxiliary equipment such as pumps, pipes, tanks, collectors, controls etc. are directly related to the type of heat source and heat sink used.

3.1 Heat Sinks and Heat Sources

Each type of source or sink has advantages and disadvantages. Table 3.0 summarizes their characteristics.

Table 3.0 - Heat Sources and Heat Sinks

	Outside Air	Exhaust Air	Ground Water	Surface Water	Waste Water	Water Loop	Earth Coils	Solar Panels
Availability	Universal	Limited	Common	Rare	Limited	By Design	Common	Universal
Suitability	Good	Limited	Good	Limited	Limited	By Design	Poor	Limited
Development Costs	None	Moderate to High	High	Low to Moderate	Moderate to High	Moderate to High	High	High
Operating Costs	Low to Moderate	Low (Reclaim)	Low	Low	Low (Reclaim)	Low (Reclaim)	Low to Moderate	Low
Aux. Heat Required	Depends on Location	Depends on Design	Depends on Loc/Design	Depends on Loc/Design	Depends on Design	Usually	Depends on Location	Usually
Temperature Limits	-10 and up	50 and up	40°F & up	40°F & up	40°F & up	60°F to 90°F	40°F & up	90°F to 200°F
Flow Rates	N.A.	by Design	1 to 5 GPM/ 12000 BTUH	1 to 5 GPM/ 12000 BTUH	by Design	by Design	by Design	by Design
Design Information	Plentiful	Scarce	Adequate	Adequate	Scarce	Available	Poor	Available
Consider-ations	Requires Defrost	Requires Engineering	Maintain Water Side	Water PH, Freezing	Requires Engineering, Corrosion, Scale	Considerable Engineering	Repairs to Buried Pipe, Design Uncertainty	Considerable Engineering
Cost Effectiveness	Good to Excellent	Excellent by Design	Depends on Alternatives	Depends on Alternatives	Excellent by Design	Depends on Alternatives	Questionable	Poor
Auxiliary Equipment	None	Ducts/Fans Dampers Filters Controls	Pumps/Pipes Tanks/ Valves Filters Exchanger	Pumps/Pipes Tanks/ Valves Filters Evaporative	Pump/Pipe Tank/ Valves Heat Valves Controls	Pipe/Pump Tank/ Valves Boiler Cooler Controls	Pipe/Pump Tanks/ Valves Controls	Collectors Pipe Tanks Pumps/ Controls

3.2 Refrigeration Cycle

Refrigeration cycle theory can be found in numerous service manuals and text books. Since it is not necessary to use refrigeration cycle theory for design and specification work, this manual will only summarize some basic refrigerant side principles. However, one important thing to keep in mind is that the operating temperature of the external evaporator/condenser coil varies considerably. For example, the outdoor coil of the air to air heat pump may operate as an evaporator in below zero temperatures and as a condenser at temperatures over 100°F. (One big advantage of the water source heat pump is that the external evaporator/condenser usually operates between 40°F and 70°F, and often operates at almost constant temperature.) This is important because temperature extremes are hard on machinery and low evaporator temperatures or high condensing temperatures reduce efficiency. Wide variations in operating temperatures also affect the heat pump refrigeration cycle charge requirements and flow characteristics. Figure 3.0 illustrates the major components of the heat pump refrigeration cycle which are discussed below.

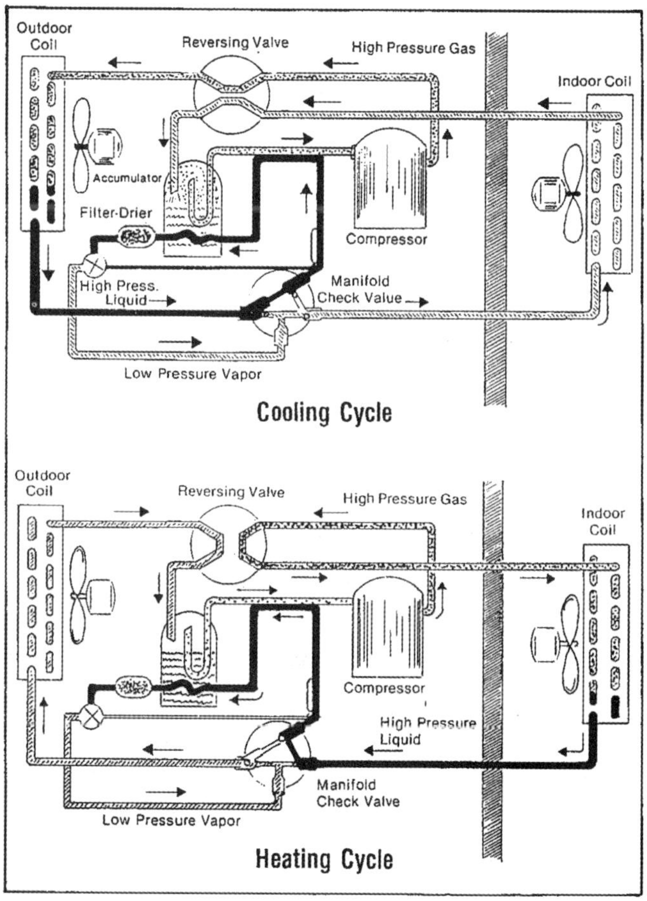

Figure 3.0 Components of the Heat Pump Refrigeration Cycle

Compressor - The compressor performs the same function in either the heating or cooling mode. Its function is to transform low pressure, low temperature gas into high pressure, high temperature gas.

Condenser - The high temperature gas is transformed into a warm liquid in the condenser. The heat that is released by the condensation process must be removed by the condensing fluid. Finned refrigerant coils are required if air is used to remove the heat of condensation, and shell and tube heat exchangers are used if the heat is rejected to a liquid. During the heat pump cooling mode, the heat of condensation is rejected to a heat sink located outdoors (atmosphere, return well, etc.) and during the heating mode, the heat of condensation is released indoors and used to heat the structure.

Refrigerant Metering Device - This device can be an orifice, capillary or expansion valve. It controls the flow of refrigerant in response to the load on the evaporator. Capillary tubes and orifices are limited in their ability to adjust to changing loads and must be selected and sized with great care. Expansion valves are more costly than capillary tubes but they are able to exercise control over a much wider range of operating conditions.

Evaporator - The liquid refrigerant admitted to the evaporator absorbs heat as it changes from a liquid to a gas. Finned coils are used to capture heat from air. Tube in tube, or shell and tube, heat exchangers are used when the heat source is a liquid. During cooling, the evaporator removes heat from the structure and in the heating mode, the evaporator captures heat from an external heat source.

Reversing Valve - When the heat pump switches modes of operation, the condenser and evaporator exchange functions as the reversing (four way) valve switches the direction of refrigerant flow through these devices. For example, during cooling, the air to air heat pump outdoor coil is the condenser. But, during heating, it becomes the evaporator. Likewise, the indoor coil is the evaporator during cooling and it acts as the condenser in the heating mode. (Because both the indoor and the outdoor coil will be required to function as an evaporator, each coil is equipped with its own metering device and bypass valve.) The reversing valve acts in response to a call for heating or cooling by the room thermostat.

Accumulator - During heating, liquid can collect in the outdoor coil (acting as an evaporator). The accumulator protects the compressor if the refrigerant floods this coil and begins to return to the compressor. The accumulator also collects any liquid carry over that is produced during the defrost cycle when the outdoor coil acts as a condenser.

3.3 Refrigerant Side Accessories

In addition to the components described above, accessories such as strainers, dryers, sight glasses, charging valves and receivers protect the equipment or aid in servicing the equipment.

3.4 Refrigerant Side Safety Devices

Crankcase Heaters - When the compressor is off, refrigerant tends to migrate to the coldest spot in the system. If the condensing unit is installed outdoors, the refrigerant will migrate to the compressor and mix with the crankcase oil. This can result in low oil levels because the refrigerant will carry the oil away from the crankcase. Another problem is liquid slugging which results from the tendency of the oil/refrigerant mixture to foam. Crankcase heat will prevent migration from occurring when the compressor is not running.

High Pressure Control - High ambient, dirty coils or filters, low air flow, incorrect refrigerant charge, or a contaminated charge can result in excessive head pressures. The high pressure control prevents oil breakdown and/or compressor damage that will occur if the unit is exposed to excessive head pressures and temperatures.

Low Pressure Control - This device protects against evaporator freeze up and/or liquid slugging by the compressor. Reduced efficiency can also be caused by low suction pressures. Dirty coils or filters, improper charge, low charge, or contaminated charge are faults that can cause the low pressure condition.

Motor Overload Protection - This device senses motor temperature and/or current draw and interrupts operation if temperature or amperage limits are exceeded.

Low Ambient Lockout - This device prevents compressor operation if the outdoor temperature falls below what would be considered a safe operating temperature. Low ambient operation can cause high compression ratios and low efficiencies. In the heating mode, modern heat pumps are capable of operating in outdoor ambients which are below 0°F. In the cooling mode, a provision for low ambient operation is required if the unit must provide mechanical cooling at relatively cold outdoor temperatures.

Anti-Cycling Device - Continuous cycling on one of the operating or safety controls can burn out the compressor motor and contaminate the refrigerant system. Small, single phase compressors are especially vulnerable because they will stall if they are started against the head pressure that was developed during the previous operating cycle. The anti-cycling device prevents compressor short cycling. Some control circuits prevent an automatic restart if the unit is shut down by a safety control and some do not.

Low Water Temperature Cutout (Freeze Stat) - This device protects against low evaporator temperatures when a heat pump is used to extract heat from water for heating or to produce cold water for cooling.

3.5 Capacity Control

Residential heat pumps are usually equipped with a single compressor that has no provision for adjusting its capacity to the actual load. At part load, the unit must cycle on and off to maintain space temperature. This causes wear and strain on the equipment, loss in efficiency, and reduced temperature and humidity control. Unless the cooling load is greater than or nearly equal to the heating load, the equipment size may be a compromise between cooling performance and heating efficiency.

In colder climates, full advantage of the heat pump's efficiency cannot be utilized because the unit size is limited by the smaller cooling load. (Refer to Section 1.5 for a complete discussion.) One solution to this problem is to install a heat pump equipped with dual compressors or a dual speed compressor. This equipment can be sized so that full capacity is used to meet the heating requirement with reduced capacity for cooling. If the heat pump is equipped with dual or two speed compressors, the evaporator coils should be split so that the coil capacity can always be matched to the compressor capacity.

Dual or varying speed equipment also has advantages in warm climates, but the situation is reversed. In this case, the design cooling load would require operation at full capacity and the less severe heating load could be satisfied by operating at reduced capacity. Figure 3.1 illustrates the multiple balance points associated with the use of dual capacity equipment.

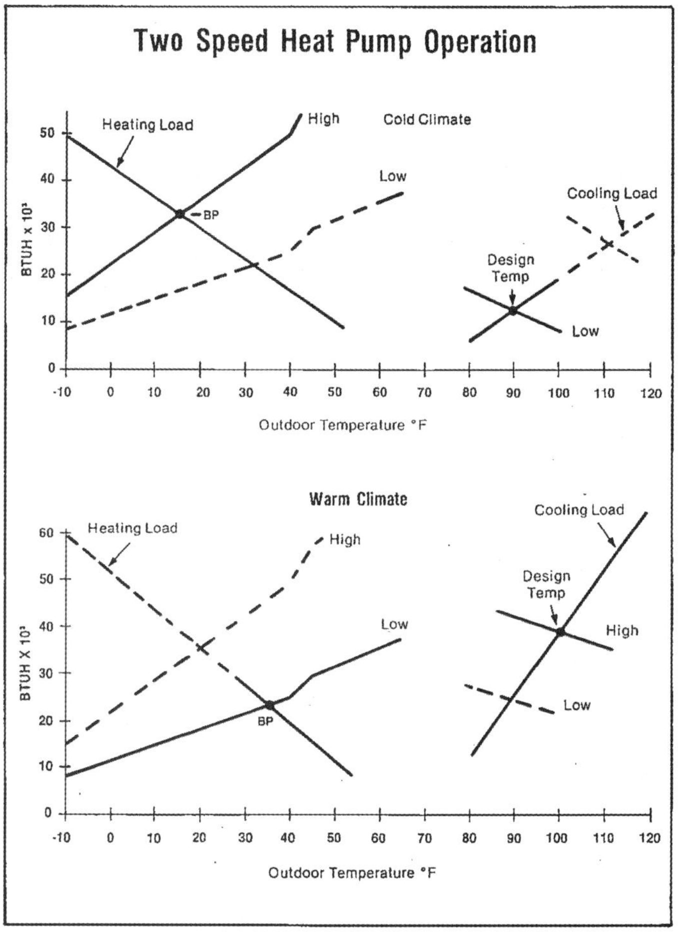

Figure 3.1 Balance Point Diagram for Two Speed Heat Pump

The ideal solution to the problem associated with unfavorable heating to cooling load ratios is to use a compressor that can vary its speed in accordance with the load. The principle is the same as explained for the dual speed unit except this unit could adjust to the exact load and cycling would be eliminated. Speed control will also allow the unit to track changes in the cooling or heating loads. This method of capacity control is included in the product line of some major manufacturers.

Commercial equipment over five tons is normally available with some form of capacity control. Multiple compressors, cylinder unloaders, hot gas by pass and split coils are either standard or accessory equipment. It is common practice to use capacity control on commercial work because the additional cost associated with these devices is more than offset by the improvement in part load performance, comfort and overall efficiency.

3.6 Humidity Control During the Cooling Cycle

In either commercial or residential work, make sure that the unit has enough latent capacity to control the space humidity. Units that are oversized, or have a coil which is too large for the compressor, should be avoided. Units installed in colder climates can be oversized on cooling by as much as to 25 percent without problems. Units installed in warm, humid climates (parts of Florida, Texas and the Gulf Coast) must be carefully selected to insure that they have adequate latent capacity. Note that heat pumps with very high EER or SEER ratings may not have enough latent capacity if they are used in warm, humid areas because these units are equipped with large evaporator coils which increase efficiency and sensible capacity but reduce latent capacity.

Usually, two speed or variable speed fans can improve humidity control during the cooling cycle. As fan speed is reduced, the velocity of the air moving across the coil is reduced and latent capacity is increased. However, the designer must make sure that the space sensible loads are satisfied and that acceptable air distribution is maintained at the lower CFM.

3.7 Defrost Cycles

Frost is always a consideration if air source equipment is required to operate when the outdoor temperature falls below 45°F. Under these conditions, the coil surface temperature will be cold enough to condense and freeze the moisture in the air. Defrost cycles are not required on water source or reclaim air source equipment.

The rate of frost accumulation depends on the outdoor temperature, relative humidity, precipitation and wind velocity. The worst frosting conditions occur during periods of high humidity when the outdoor temperature is between 30°F and 40°F. Frost accumulation is less severe at colder temperatures because cold air does not contain much moisture. Figure 3.2 illustrates how humidity and temperature affect the need for heat pump defrosting. The number of defrost cycles depend on how many days per year the various temperatures occur and this will vary greatly from region to region.

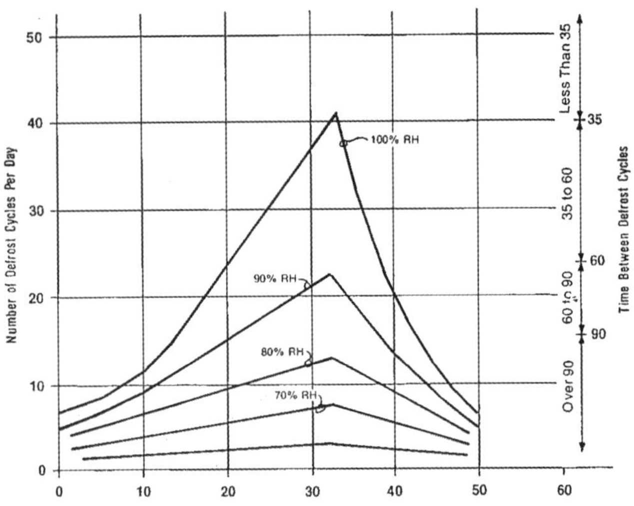

Figure 3.2 Defrost Requirements

Laboratory research has shown that overall unit efficiency is maximized if the defrost cycle is initiated when the coil is 50 to 60 percent frosted. Lower accumulations do not degrade the heating performance enough to justify the using energy expended by the defrost cycle. The ideal defrost cycle would sense the amount of frost accumulation and initiate defrost only as required.

"Reverse cycle" defrosting is by far the most common method of defrost. In this method, the unit is switched to the cooling mode and defrost occurs as hot gas from the compressor condenses in the outdoor coil. During defrost, the outdoor fan is usually de-energized because it would work against the defrosting process. This method requires the use of auxiliary resistance heaters. This is because during defrost, the unit is trying to cool the space and the auxiliary heat must be activated to temper the cool supply air. Ideally, the number of defrost cycles should be held to a minimum because the compressor is subjected to wear and strain every time defrost is initiated. Timed defrost and demand defrost are the two basic methods of initiating defrost.

Timed defrost can be set to initiate the defrost cycle every 30, 60 or 90 minutes depending on the local weather conditions. But, this method allows the unit to enter the defrost cycle even if there is no frost on the coil. To overcome this problem, the times defrost cycle usually includes a provision for overriding the defrost cycle if the coil temperature is above freezing. This method is called timed-temperature defrost and it significantly reduces the number of unnecessary defrost cycles. This system is relatively foolproof because the time and temperature input can be accurately measured under all operating conditions. One drawback to the time-temperature method is that it cannot prevent unnecessary defrost cycles during periods of cold, dry weather. Another disadvantage is that the timer setting must be selected for the most severe condition, even though this condition occurs only a small percentage of the time.

Demand defrost is the second method of initiating defrost. This method will not activate the defrost cycle unless the coil has reached the 50 to 60 percent frost condition. Unnecessary defrost cycles are eliminated and the overall system efficiency is improved. The drawback to this method is that it is not easy to determine when the coil is at the required frost condition because extraneous factors such as wind, precipitation, humidity, snow, leaves or poor maintenance can interfere with the defrost control. Presently, there are two techniques that are used for initiating demand defrost.

1) The first technique measures the outdoor ambient temperature and the refrigerant temperature at the outdoor coil. The difference between these temperatures increases as frost accumulates on the coil. Defrost is initiated when this temperature difference reaches a predetermined value. The problem with this method is that the variation in temperature difference (control signal) is relatively small and changes in outdoor humidity, space loads, or refrigerant charge can affect the signal.

2) The second technique measures the air pressure drop across the outdoor coil. Defrost is initiated when this difference reaches a predetermined value. Unfortunately, outside factors such as wind, leaves, debris, or a dirty coil can produce a false signal. However, a well designed control will not initiate defrost unless the pressure signal is steadily maintained for some reasonable interval of time and is confirmed by a coil temperature reading.

Both the demand and the time-temperature defrost systems use the temperature of the refrigerant leaving the outdoor coil as the termination signal. This signal is reliable because this temperature will rise rather dramatically after the frost has been dissipated. Well designed systems also include a back up termination method which consists of a timer that interrupts the defrost cycle after a preset time. This device will also terminate defrost if the primary control malfunctions, or if the frost build-up is too severe to dissipate in a reasonable amount of time.

Because the defrost cycle is a parasitic load, the seasonal efficiency of the heat pump is degraded by initiating unnecessary defrost cycles. The table below shows how dramatically the method of defrost control can affect the annual number of defrost cycles.

Number of Annual Cycles

Location		Time/Temp. Defrost 30 Min.	60 Min.	90 Min.	Demand Defrost
1 York	PA	3592	1796	1197	178
2 Tacoma	WA	2372	1186	790	112
3 Charlotte	NC	2145	1072	715	253
4 Dallas	TX	1022	511	340	177
5 Atlanta	GA	2364	1182	788	78
6 Ithica	NY	5972	2986	1990	205
7 St. Louis	MO	3183	1591	1061	396

(Extracted from Manufacturers Test Data)

3.8 Microprocessor Controls

A microprocessor-based control can be used to perform all the safety functions listed in Section 3.4 and defrost demand control. The microprocessor constantly monitors the pertinent temperatures and pressures and compares these values with operating information stored in its memory. This information, and the logic that is preprogrammed into the device, allows the microprocessor to make the necessary control decisions. One advantage of the microprocessor is that only one device is required to perform all the various control functions. (The number of sensors is not reduced.) Another advantage is that the microprocessor can be programmed to cross check the various signals and verify any condition. The device can also be programmed to differentiate between a substandard operating condition and a dangerous fault. It can also alert the owner of the condition or automatically shut the unit down. The microprocessor is usually compatible with a plug in service analyzer which allows the serviceman to monitor all the system parameters during startup or when he is trouble shooting.

3.9 Efficiency Ratings

Various methods are used to test heat pump performance. Efficiency information must be determined for both the heating and cooling modes. In the heating mode, the COP and the HSPF are used to evaluate performance. Cooling performance factors consist of the EER and SEER ratings.

The coefficient of performance (COP) is an indication of heat pump efficiency when it operates continuously under a steady load. The COP is calculated by dividing the heat output of the unit (in Btuh) by the power consumed by the unit which includes fan power (in Btuh).

$$COP = \frac{\text{Steady State Heat Output (Btuh)}}{\text{Steady State Power Consumed (Btuh)}}$$

COP tests are based on the methods developed by the American Refrigeration Institute (ARI). The ARI standard determines the steady state COP for evaporator (outdoor ambient) temperatures of 17°F and 47°F for air source equipment or 50°F and 70°F for water source equipment. All the tests are conducted at a space temperature of 70°F.

The tested COP can be a very misleading indication of heating season efficiency because it does not account for many factors that affect heat pump performance.

- Air source heat pump performance constantly changes as the outdoor temperature changes.
- The affects of supplemental heat are not included.
- The affects of changes in indoor temperature and internal loads are not included.
- The affects of unit on/off cycles are not accounted for.
- The affects of defrost cycles are not accounted for.

The heating season performance factor (HSPF) is another method used to indicate heating efficiency. The HSPF is determined from a test that attempts to account for the effects of on/off cycling, varying source temperatures and parasitic loads. The HSPF is equal to the total estimated heat output from the unit for an entire heating season (in Btu) divided by the total estimated energy consumed (in Btu) by the unit during that season.

$$HSPF = \frac{\text{Estimated Total Heat Energy Delivered (Btu)}}{\text{Estimated Total Energy Consumed (Btu)}}$$

Once again, complications arise because the number and duration of the unit operating cycles, defrost cycles and supplemental heat cycles will vary with local weather patterns. In other words, the HSPF of any given unit will depend on where it is installed and how it is operated. In an effort to compensate for local weather patterns, the HSPF is determined for the six different regions shown in Figure 3.3 (Note that the HSPF for Region IV has been selected by the Federal Trade Commission for the purposes of advertising and publication. Also note that the HSPF does not apply to "add on" heat pump applications.) While the HSPF test is a better indication of heat pump efficiency than the COP test, it still does not fully account for the factors listed above.

Regional Heating Load Hours

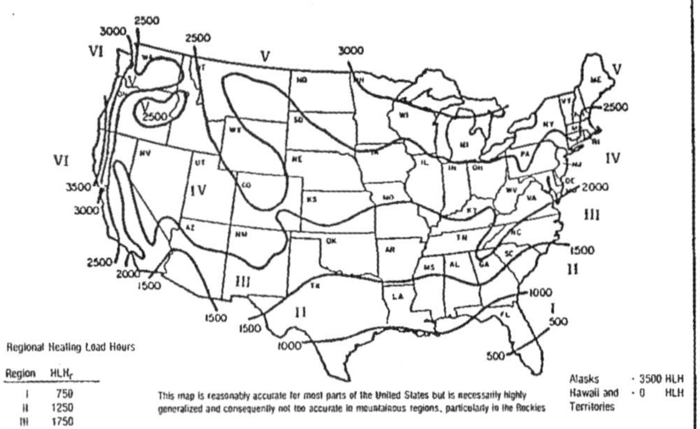

Heating Load Hours are the Number of Hours the Equipment Would Have to Operate at Full Capacity so That the Total Heat Supplied Under Full Load Operation Equals the Total Amount of Heat Required for the Entire Heating Season.

Figure 3.3 Regional HSPF

The published COP or HSPF information is only a relative indication of unit efficiency and should only be used for comparisons of similar pieces of equipment. A more accurate method of estimating energy consumption for a specific application is provided by the bin calculation procedure. Until recently, these calculations were too time consuming to be performed by hand, but now microcomputers and technical software allow these calculations to be made in a matter of minutes. Note that the results of the bin calculation may vary from actual field measurements because the accuracy of the bin calculation is closely related to the accuracy of the load calculations. Credit for internal loads and solar loads should be taken when a bin calculation is used to estimate heating energy. Some utilities have field test information which may be helpful when making estimates for heating energy consumption.

In the cooling mode, the energy efficiency ratio (EER) is used to indicate unit performance. The EER is calculated by dividing the steady state cooling capacity (in Btuh) by the unit power consumption (in watts). The EER rating is determined for a specific set of operating conditions outlined in the ARI standards. (Air cooled equipment is tested with 95°F air entering the condensor. Water cooled equipment is tested with 70°F water entering the condensor at the flow rate specified by the manufacturer. The space temperature is 80°F D.B. and 67°F W.B. in both cases.) However, the EER rating does not account for the following factors:

- The cooling performance constantly changes as the condensing temperature changes.
- The effects of changes in indoor temperature, solar gains, and internal loads are not included.
- The effects of part load operation (unit on/off) cycles are not accounted for.

In an attempt to account for the factors listed above, a seasonal energy efficiency ratio (SEER) has been developed. The SEER rating test attempts to estimate the actual cooling season performance by including some of the effects of part load operation. The SEER is calculated by dividing the total estimated cooling delivered (in Btu) for a cooling season by the total estimated energy consumed (in watt hrs) for that season.

$$\text{SEER} = \frac{\text{Estimated Cooling Delivered (Btu)}}{\text{Estimated Energy Consumed (watt hrs)}}$$

Note that the SEER rating is determined from tests conducted at specified part load conditions and does not take all variations of weather and load conditions, or changes in condensing and evaporator temperatures, into account. However, the SEER value is useful for comparing the cooling performance of similar pieces of equipment. The total cooling season energy consumption of a specific piece of equipment at a specific location can be estimated by the bin calculation method. In addition, local utilities may have information that may be helpful in making cooling energy consumption estimates.

3.10 Strategies for Control of Heat Pump Operation

Operating controls play a major role in achieving efficient operation. On any heat pump system, the preferred control strategy requires the use of the most economical heat source to the greatest possible extent while minimizing the use of alternate heat sources. The designer must consider equipment efficiency, local fuel costs, and local weather conditions to determine the most economical heat source at any given outdoor temperature. In most applications, the most economical control strategy is usually obvious:

1. Air to air heat pump with supplementary resistance heat or heat pump added to an electric furnace.
   - Heat pump should operate continuously below the balance point.
   - Electrical resistance heat should be locked out above the balance point and brought on line in stages as the outdoor temperature drops. While staging does not necessarily reduce energy consumption, it does limit electrical demand and in some applications staging will result in a lower electric bill. In addition, staging improves comfort because the heat delivered is always close to the amount of heat required.

2. Air to air heat pump and fossil fuel furnace.

   In this application, the economic balance point must be determined before any control decisions are made. The economic balance point is equal to the outdoor temperature at which it becomes less expensive to heat with fossil fuel. The economic balance point occurs because the heat pump COP decreases as the outdoor temperature decreases. Thus, the efficiency of the fossil fuel furnace increases as full load conditions are approached. Equipment performance and efficiencies, fuel costs and electrical rates will determine the economic balance point. If fossil fuel is relatively inexpensive, the economic balance point will be above the thermal balance point. If electricity is relatively inexpensive, the economic balance point will fall below the thermal balance point. Refer to Section 2.6 for a more thorough discussion of the economic balance point.

   The furnace should cycle under control of the space thermostat when the outdoor temperature is below the economic balance point. The heat pump should operate when the outdoor temperature is above the economic balance point. If the furnace is not heating, it is acceptable for the heat pump to operate when the outdoor temperature is below the economic balance point.

   The heat pump should not operate when the furnace is heating because manufacturer warranties and some building codes may be violated. And, if the indoor coil is located in the duct downstream from the furnace, simultaneous operation will result in high head pressures and possible compressor damage. (Refer to the last paragraph in Section 4.3 and Section 5.13 for a discussion regarding the location of the "add on" indoor coil.)

3. Ground water heat pumps (water to air).

   Ground water heat pump systems are available with mechanical cooling or direct water coil cooling. (Refer to Section 2.3 for a discussion of both designs.)

   A ground water system that cools mechanically must be selected to satisfy both the heating and cooling requirements of the space. In this case, good cooling performance requires that the cooling capacity not exceed the design cooling load by more than 25 percent. Supplementary heat may be required and the system balance point will be determined by the cooling load limitation. If electric resistance coils are used for supplemental heat, the control strategy will be the same as discussed for the all electric air to air heat pumps. If a fossil fuel furnace is used for supplemental heat, the control strategy is the same as the one used for air to air heat pumps.

   A ground water system that is equipped with water coil cooling can be sized so that it provides 75 to 100 percent of the heat required by the structure and little, or no supplementary heat will be necessary. If no supplementary heat is used, the heat pump simply operates to satisfy the space thermostat. If supplementary heat is used, the heat pump should continue to operate when the supplementary heat is energized.

   All ground water systems require controls for the water supply system. These controls should include provisions for:

   - Maintaining refrigerant pressures by controlling the water flow when the system is operating.
   - Preventing water flow when the system is not operating or is shut down by a refrigerant side safety control.
   - Cycling the water pump to maintain water pressure in the pressure tank.

4. Air to water heat pump and hot water boiler.

   The air to water heat pump could be used as an "add on" unit to an existing hydronic system. This application is similar to the fossil fuel furnace "add on" application because the control strategy will depend on both the thermal balance point and the economic balance point. Above the change-over (thermal or economic) balance point, the heat pump should operate and supply hot water (90°F to 135°F) to the hydronic heating system. Below the change-over balance point, the boiler should supply the required hot water (135°F and higher) and the heat pump should not operate. The controls should also prevent heat pump operation when the temperature of the water circulating through the heat pump is above the safe operating temperature as specified by the manufacturer.

Heat pump operation can be controlled directly from the space thermostat or indirectly by an aquastat in the hot water supply system. Changeover from the heat pump to the boiler should occur when the outdoor temperature is below the changeover balance point and the second stage of the space thermostat is calling for heat.

It is also possible to use zone control with this system by allowing the first stage of a space thermostat to control a zone valve which in turn can call for operation of the heat pump. The second stage of the room thermostat can be used to stop the heat pump and to activate the boiler.

Some units are capable of providing domestic hot water as well as hot water for hydronic heating. Refer to the manufacturers recommendations for details on controls and piping.

5. Ground water to water heat pump and hot water boiler.

   The comments made for the air to water heat pump and the hot water boiler also apply to this equipment. In addition, the water source controls that are listed for the ground water heat pump are also required in this application.

6. Closed water loop heat pump systems (reclaim).

   These systems require controls that operate the individual heating and cooling units, the evaporative cooler, the boiler and, in some cases, a thermal storage system and/or an economizer cycle. The details for controlling this type of system is available from the firms that manufacture equipment for this application. Refer to Section 2.8 for more information on how these systems operate.

7. Commercial heat pump applications.

   Commercial systems may include provisions for simultaneous heating and cooling, economizer cycles, day/night cycles, variable air flow rates, supply air set point temperature schedules and minimum ventilation requirements. No single control strategy is applicable to every situation. While some applications are straight forward, some are not. Some engineering time is necessary to integrate the controls and the equipment into an energy efficient system.

8. Reclaim heat pump applications.

   The heat that is available for reclaim seldom equals the heat required by the structure. The controls must decide to store or expel excess reclaimed heat and must energize supplementary heat when the quantity of reclaimed heat is not adequate. In some applications, the system operating temperatures and efficiencies vary, and controls may be required to select the most efficient operating mode. The controls for systems that reclaim heat from exhaust air, or liquid wastes must be tailored to the needs of each specific application.

9. Water heating heat pumps.

   The specifications for the operating and safety controls for these units are usually available from the manufacturer. No special problems arise if water to water units are used to extract heat from ground water or waste fluids, or if air to water units are used to extract heat from outdoor air or exhaust air. Refer to Section 2.12 for additional information.

3.11 Operating Controls

Space Thermostat - This thermostat is usually a two stage device. Above the balance point, the first stage cycles the compressor and below the balance point, the second stage cycles the auxiliary heat. During cooling, the thermostat cycles the compressor. Changeover between heating and cooling can be automatic or manual, and most thermostats are equipped with a manual fan switch. If a dual compressor or a dual speed heat pump is used, a three stage space thermostat can be used to bring on compressor capacity in stages and energize the auxiliary heat as required.

Outdoor Thermostat - Outdoor thermostats can be used to lock out the auxiliary heat when the outdoor temperature is above the balance point. If no outdoor thermostat is used, the auxiliary heat can be activated at any outdoor temperature by abruptly raising the setting of the room thermostat. This will lead to a faster heating response, but the electrical demand load may be increased unnecessarily. Energy consumption will also be increased because auxiliary heat (COP of 1) will be used instead of the heat pump (COP greater than 1). The absence of an outdoor thermostat will also allow the resistance heaters to compensate for a heat pump that is not operating up to its capacity and energy consumption will be increased.

If the auxiliary heat is to be brought on line in stages, each stage will require a separate outdoor thermostat. This type of control is usually used to keep the electrical demand load to a minimum. The extra thermostats required for "staging" can be cost effective in situations where demand billing is in effect. Figure 3.4 illustrates how the outdoor thermostat setpoints can be determined from the balance point diagram.

Note: Heat pumps that are equipped with a reverse cycle defrost system, will require a control circuit that allows the defrost control to energize the auxiliary heat at any outdoor temperature, even if it is above the balance point. If the auxiliary heat is "staged", only part of the installed auxiliary capacity may be required.

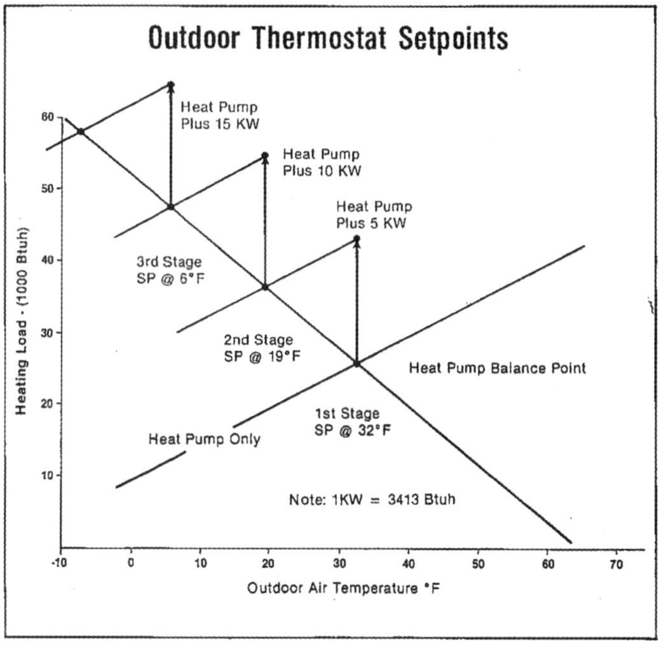

Figure 3.4 Outdoor Thermostat Balance Points

Programmable Setback Thermostats - These thermostats are designed with a microprocessor unit which allows the thermostat to be programmed for one or more "setback" periods per day. During recovery, the microprocessor unit monitors the space temperature and adjusts the heat pump performance so that the space temperature is raised in a gradual manner. Recovery is achieved over an extended period of time and is accomplished with maximum use of the heat pump and minimum use of auxiliary heat. Owner operating costs are reduced as a result of the reduction in auxiliary heat operating time. This thermostat also can be used to provide setback for an add-on heat pump, and it can help save cooling energy by programming it to increase the cooling setpoint when the space is vacant.

Note: This device may not be effective on heat pumps that are sized correctly and are equipped with an outdoor thermostat. (When the outdoor temperature is above the balance point, a properly adjusted outdoor thermostat will not allow the auxiliary heat to be energized during recovery from manual setback. When the outdoor temperature is below the balance point recovery always requires the use of the auxiliary heat.)

Aquastats - Aquastats are used for controlling water temperatures on the heating side of an air to water, or water to water system. They can be used to perform operating or safety control functions. Aquastats are not normally used to control room temperature because they do not directly measure the conditions in the space.

Section 4

Heat Pump System Application Principles

All heat pump systems should be engineered and designed to meet the specific requirements of any given project. Sometimes the calculations, plans and specifications are quite simple and at other times they are quite extensive. However, in every case, the design procedure must follow the same basic pattern.

- Determine the heating and cooling loads.
- Select equipment.
- Select and locate air terminals.
- Meet outdoor air requirements.
- Size and locate ducts and pipes.
- Locate all equipment.
- Establish and meet noise criteria.
- Meet specified or recommended installation requirements.
- Select controls.
- Evaluate first cost and operating cost.

Knowledge of certain design principles is required to do the work outlined above. Many of these principles are common to other types of heating and cooling systems, but some principles apply only to heat pumps. This section will discuss these design principles with special emphasis on those items that are peculiar to heat pumps.

4.1 Load Calculations

Load calculations must be made for heat pump systems, or any other type of heating and/or cooling system. The importance of an accurate load calculation cannot be overemphasized. It is the source for all of the system design parameters. Load calculations fall into three distinct categories:

Residential - The methods described in the latest edition of ACCA's Manual J should be used to estimate residential loads.

Small Commercial (20 Tons and Less) - Refer to the current edition of ACCA's Manual N for the correct load calculation procedure.

Large Commercial and Industrial - The ASHRAE GRP 158 Manual should be used in this case.

There is a fourth category, "process loads", that is not covered by any of these manuals. Process loads usually must be evaluated on an individual basis and require the services of a qualified engineer. Process loads are encountered in comfort conditioning and heat recovery work.

Load calculations should be used for more than determining the maximum (design) heating and cooling loads. For example, systems that include a provision for capacity control require an estimate of the minimum cooling load and changeover control decisions require load estimates at "off design" conditions.

Estimating loads for all operating conditions is especially important for heat pump system design work. These systems require knowledge of the loads at any outdoor temperature so that the thermal balance point and the supplementary heat requirements can be determined. Section 1.6 of this manual illustrates how residential and commercial loads vary as the outdoor temperature varies. Estimating the loads at any outdoor temperature is relatively simple for residential buildings, but considerable skill may be needed to evaluate commercial and industrial load characteristics.

4.2 Air Distribution

Air distribution means selecting, sizing, and locating the supply and return registers. It is extremely important that this task is done correctly because it has a direct effect on the comfort of the occupants and the performance of the system. Complaints of poor heat pump performance can often be traced to poor air distribution. A good air distribution system will introduce air into the occupied space without creating drafts or stratified zones and will maintain a relatively constant temperature and air movement throughout the space.

Each separate space requires a certain quantity of supply air. The exact supply CFM for any given space will be determined by either the cooling or the heating requirement. A load calculation for each individual space or zone is required so that the design CFM can be estimated before the air terminals are selected.

The CFM required for each space or zone can be estimated by the following formula.

$$\text{CFM} = \frac{\text{Btuh}}{1.1 \times \text{T.D.}}$$

Where:

CFM = Room or Zone CFM
Btuh = Room or Zone Sensible Load
1.1 is the Air Constant
T.D. is the approximate difference between the room temperature and the supply air temperature.

Example - Determine the room CFM

Sensible room cooling load = 2,500 Btuh

Room heating load = 4,000 Btuh

Cooling supply air = 58°F
Heating supply air = 105°F

Room temperature .... cooling = 75°F

Room temperature .... heating = 70°F

$$\text{Cooling CFM} = \frac{2{,}500}{1.1 \text{ X } (75\text{-}58)} = 134 \text{ CFM}$$

$$\text{Heating CFM} = \frac{4{,}000}{1.1 \text{ X } (105\text{-}70)} = 103 \text{ CFM}$$

In this case, 134 CFM is the approximate design CFM. Notice that knowledge of the supply temperatures is necessary for this calculation. If a specific piece of equipment has been selected, the supply temperature information is found in the manufacturers data. If the equipment has not been selected, the designer can assume the supply temperatures based on previous experience with the type of equipment under consideration. (In this case, the designer is obligated to verify the assumptions after the equipment has been selected.)

For residential work, the methods and procedures outlined in ACCA's Manual D, and on the D-1 and D-2 forms should be used to determine the room CFM.

Once the room or zone CFM is known, the air terminals can be sized, selected and located. Refer to ACCA's Manual E for a discussion on air terminal performance. Residential air distribution is also covered in, ACCA Manuals D and G. Commercial air distribution is covered in ACCA Manual Q. In all cases, manufacturers catalog data is necessary to complete the air distribution system design. Table 4.0 summarizes some of the guidelines discussed in the references.

Table 4.0
Location of Supply Terminals

Type	Location	Discharge	Throw	Performance in Warm and Cold Climates.
Floor Baseboard	Perimeter	Vertical Spreading	to Ceiling	Good Performance in all Applications.
Floor Baseboard	Perimeter	Vertical Non Spreading	to Ceiling	Acceptable Performance in all Applications.
Baseboard Low Sidewall	Perimeter	Horizontal or Inclined	Across Room	Air Stream Directed at Occupants, Not Recommended For Heat Pump or Cooling Applications. Produces Stagnant Area in Occupied Zone When Used for Cooling.
High Sidewall	Interior	Horizontal	to Outside Wall	This System Works Well in Warm Climates and Can be Used in Tight, Well Insulated Structures in Cold Climates. Not Recommended if Floors are Exposed to Cold Temperatures.
Low Sidewall	Interior	Horizontal or Inclined	Across Room	Air Stream Directed at Occupants, Not Recommended for Heat Pump or Cooling Applications.
Baseboard Low	Interior	Vertical -Steeply Inclined	Up Into Room	This System Will Work in Warm Climates and in Tight, Well Insulated Structures in Cold Climates.
Ceiling	Perimeter	Vertical Spreading	to Floor	This System Works Well in Warm Climates and Can be Used in Cold Climates if the Floors are Not Exposed to Cold Temperatures.
Ceiling	Ceiling Pattern	Horizontal	To Walls or to Adjacent Air Pattern	This System Works Well in Warm Climates and for all Cooling Applications. It can be Used for Heating in Cold Climates if the Floors are Not Exposed to Cold Temperatures.

One important point to remember about heat pump systems is that the temperature of the supply air during the heating cycle (without auxiliary heat) may be only 85°F to 100°F and it can feel relatively cool as it comes out of the register. This characteristic makes the terminal selection and location even more critical.

Return registers are also important. Returns must be sized and located so that the air has an unobstructed path back to the air handler. This means that when a single central return is used, the doors will have to be undercut or vented. Otherwise, multiple returns are required. Returns do not have much effect on room air patterns. Therefore, the exact location of returns is not critical. However, care must be taken to avoid short circuiting the supply air directly to a return. Note that a well-designed return system cannot compensate for a poorly designed supply system. But, a poorly designed return system can degrade the performance of a well-designed supply system. The references noted above also discuss the sizing, selection and location of returns.

If a heat pump is added to an existing warm air system, the existing air terminals may be improperly sized for the air quantities that are required for the heat pump. In some cases, new terminals must be selected and installed.

Even a perfectly designed distribution system may not perform up to expectations unless it is balanced after the system is installed. The air balance is a final and necessary step that is required to insure proper air distribution. For this reason, hand dampers should be included in the duct branches to facilitate balancing. Finally, all of these good efforts may be ruined if the occupants block off some air terminals with furniture or draperies.

4.3 Duct Design

The heat pump, the duct system and the air terminals must work together. Each component must be designed to do its job or the system will perform poorly. It is not unusual to find blame for poor heating or cooling performance directed toward the heat pump when the real problem can be found in the duct system. A well designed duct system will allow the fan to circulate the design supply and return air quantities without exceeding a predetermined static pressure loss. The duct system should not leak. The conduction of heat across the duct walls should be minimized and the duct system should not generate airborn noise.

Low velocity commercial duct systems are usually designed by the equal friction method. This method sizes the various duct sections so that they will all have the same predetermined friction loss per 100 feet of run. After all the duct runs are sized, the duct system static pressure loss is computed and a fan is selected to deliver the design CFM at the calculated static pressure. (In commercial work, the duct system is designed before the fan is selected because a wide range of fan performance options are available with commercial equipment.)

Commercial low velocity duct systems are sized so that the air velocities in the duct runs do not exceed certain recommended values. (The design friction loss per 100 feet of run usually falls between 0.08 In. W.G. and 0.20 In. W.G.) All transitions should be gradual and all fittings should be aerodynamically efficient. Refer to ACCA's Manual Q for more information on commercial duct design.

Residential duct systems should be designed by the ACCA Manual D method. This method sizes the duct system so that the supply and return static pressure losses equal the external static pressure available at the air handling unit. (In residential work, the fan is selected before the duct system is designed because residential equipment does not offer a wide range of fan performance options and the available static pressure is usually limited to only a fraction of an inch.) Refer to ACCA Manual D for more information on residential duct design.

All duct systems should be designed for quiet operation:

- Flexible connections isolate equipment vibrations.
- Duct liner attenuates airborn noise.
- 90 degrees ELS can attenuate airborn noise.
- Poorly designed ELS and fittings can generate airborn noise.
- Lined ELS are very effective at attenuating airborn noise.
- Breaks and reinforcement prevent "oil canning."

Because heat pump supply air temperatures are only 85°F to 110°F, it is especially important to insure that the duct system is thermally efficient:

- Insulate ducts located in unconditioned spaces.
- Insulate boots and fittings located in unconditioned specifications.
- Refer to codes and utility company insulation specifications.
- Seal all seams and joints.
- Vapor barriers prevent condensation within the insulation or on the duct surface.
- Refer to ACCA Manual 4 for below grade installation procedures.

Vhen a heat pump is added to an existing heating ›r cooling system, the designer must make sure hat the existing duct system and blower are adeate. If changes are required, but not made, pace conditions may be unacceptable, and low ir flow may result in compressor damage or the urnace performance may be adversly affected.

- The duct airways may not be large enough to handle the increased air flow that is required for the heat pump.
- Adding a cooling coil to an existing warm air heating system drastically increases the fan static pressure requirement.
- The existing fan may not be able to produce the static pressure required to accommodate the increase duct friction and/or coil pressure drop.
- Modifying the duct system, or changing out the fan may prove to be impractical or economically unfeasible.
- Furnace manufacturers warranties, or testing agency approvals may be voided if air flow through the furnace is altered.

.4 Water Piping

ir to water, water to water and water to air heat umps all require water piping. Water piping is reuired to connect the heat pump to the heat ource (water to air or water to water) and/or to istribute the heat to the occupied space (air to vater or water to water). The water piping systems hat can be encountered include:

- Piping between the heat pump and a well or pond.

nformation regarding this piping system is vailable in manufacturers' data, from a local well riller, or from the National Well Water Association (located in Columbus, Ohio). The design work equired for this type of system includes pump election, pipe sizing, pressure tank selection, nd the selection of safety, flow control, and hutoff valves.

- Piping between the heat pump and the heating and cooling terminals located in the occupied space.

his piping is similar to the piping systems used ith conventional chilled water cooling and hot ater heating systems. The methods and proedures used for designing this type of system an be found in the ASHRAE handbooks and in ngineering guides which are available from maor manufacturers. The design work for this type f system includes pipe sizing, pump selection, xpansion tank selection, selection of safety, ow control and shutoff valves, provisions for nakeup water, and back flow prevention.

- Piping associated with the closed loop reclaim system.

A considerable amount of information is available from the manufacturers of closed loop heat pump equipment. This information plus a knowledge of basic hydronic principles is necessary to insure a successful design.

- Piping associated with solar systems and earth coupled systems.

In this case, the designer must have a knowledge of basic hydronic and fluid flow principles. Manufacturers may provide additional design information. Information on solar design is available in recently published books and manuals.

4.5 Air to Air Equipment Selection

The designer must select equipment which provides the correct heating and cooling capacities and meets certain air flow and/or ventilation requirements. This work requires selecting the basic heat pump unit, and may include the selection of fans, fan motors, pulleys, auxiliary heaters, dampers, operating controls and safety controls.

Step 1 - Determine the approximate unit CFM.

The unit CFM will usually be determined from the sensible cooling requirement, but the heating CFM should always be checked. On "heating only" applications the heating load will determine the design CFM. In residential or commercial work, the approximate unit CFM is determined from the following formula:

$$CFM = \frac{Btuh}{1.1 \times T.D.}$$

CFM = Approximate unit CFM
Btuh = Total Space Load
1.1 is the Air Constant
T.D. is the approximate difference between the room temperature and the supply air temperature.

Example: Finding unit CFM.

Total space cooling sensible load = 25,000 Btuh

Total space heating load = 40,000 Btuh

Cooling supply air = 58°F
heating supply air = 105°F

Room temperature .... cooling = 75°F
... heating = 70°F

$$\text{Cooling CFM} = \frac{25{,}000}{1.1 \times (75\text{-}58)} = 1{,}340 \text{ CFM}$$

$$\text{Heating CFM} = \frac{40{,}000}{1.1 \times (105\text{-}70)} = 1{,}030 \text{ CFM}$$

In this case, 1,340 CFM is the approximate unit design CFM. Notice that knowledge of the supply temperatures is necessary for this calculation. If a specific piece of equipment is under consideration, the supply temperature information can be found in the manufacturers data. If the equipment has not been selected, the designer can assume supply temperatures based on previous experience. (The designer is obligated to verify these temperature assumptions after the equipment has been selected.)

Step 2 - Select a unit to satisfy the cooling requirements.

Use the manufacturers catalog data to select a unit that has a nominal CFM that is close to the calculated CFM. Use the catalog data to verify that the unit has the desired sensible and latent capacity. At this point, there are two possible situations:

1) No outdoor air is required and the unit's sensible and latent loads are approximately the same as the space loads. This is normally the case in residential work and occurs occasionally in small commercial work.

2) Outdoor air is introduced through the unit and the unit sensible and latent loads are greater than the space loads. This is common in small commercial work and always required in commercial/industrial work. Outdoor air requirements are determined from codes and/or recommendations found in the ASHRAE handbooks. (The ACCA Commercial Load Calculation Manual N includes this information but local codes have precedence.)

The sensible and latent capacity of the unit, operating at design conditions, must be equal to or greater than the calculated sensible and latent design loads.

Cooling selection requires the following information:

- Total sensible load on the cooling unit:
  Space load + outdoor air (ventilation) load
- Total latent load on the cooling unit:
  Space load + outdoor air (ventilation) load
- Condensing temperature:
  Use the same dry bulb temperature that was used for the cooling load outdoor design condition.
- Approximate design CFM:
  Determined from the calculation shown above.
- Wet bulb temperature of the air as it enters the coil:
  Determined on the psychrometric chart, given the percentage of outdoor air, outdoor design condition, and room design condition.
- Manufacturers catalog data:

Example: - Verifing cooling capacity for residential or commercial equipment.

	Sensible	Latent
Equipment Loads =	25,000 Btuh	4,000 Btuh

Approximate Unit CFM = 1,340

Condensing Temperature = 95°F

Entering Wet Bulb Temperature of air into coil = 67°F

The manufacturers data (Figure 4.1) shows that Unit 7 has the correct nominal CFM and more

Unit 6 Operating at 1300 CFM has acceptable air flow and closely matches the required sensible and latent capacity.

6 ←*

Temp (F) Air Ent Outdoor Unit		Air Ent Indoor Unit – Cfm/BF 1050/.11			1150/.12			1300/.13		
		Indoor Unit Ent Air Temp – Ewb (F)								
		72	67	62	72	67	62	72	67	62
85	TC	35.1	32.5	29.8	35.3	32.8	30.3	35.3	33.1	30.9
	SHC	18.1	23.3	27.8	18.5	24.1	28.9	18.8	25.2	30.3
	KW	4.37	4.21	4.04	4.43	4.28	4.12	4.49	4.36	4.22
95	TC	33.9	30.9	28.1	34.1	31.5	28.6	34.2	31.4	29.3
	SHC	17.6	22.8	26.9	18.1	23.7	27.9	18.8	24.9	29.1
	KW	4.64	4.43	4.24	4.70	4.50	4.32	4.76	4.58	4.43
100	TC	33.0	29.9	27.3	33.2	30.2	27.8	33.4	30.5	28.5
	SHC	17.3	22.4	26.4	17.8	23.3	27.3	18.6	24.6	28.4
	KW	4.75	4.53	4.34	4.81	4.59	4.42	4.90	4.69	4.55
105	TC	32.1	28.9	26.4	32.4	29.2	27.0	32.6	24.5	27.7
	SHC	17.0	22.0	25.8	17.5	23.0	26.7	18.3	24.3	27.6
	KW	4.87	4.63	4.44	4.93	4.70	4.53	5.02	4.79	4.65
115	TC	30.3	27.0	24.7	30.6	27.4	25.3	30.9	27.6	26.0
	SHC	16.3	21.0	24.8	17.0	22.4	25.4	17.9	23.8	26.0
	KW	5.11	4.82	4.63	5.17	4.91	4.74	5.27	4.99	4.87

Unit 7 Operating at its nominal CFM (1350) has the correct air flow and has excess sensbile and latent capacity.

7 ←

Temp (F) Air Ent Outdoor Unit		Air Ent Indoor Unit – Cfm/BF 1200/.11			1350/.12			1500/.13		
		Indoor Unit Ent Air Temp – Ewb (F)								
		72	67	62	72	67	62	72	67	62
85	TC	39.6	37.6	34.7	39.6	38.2	35.2	40.4	38.3	35.9
	SHC	20.3	26.8	32.3	20.7	28.0	33.6	21.5	29.0	35.1
	KW	5.38	5.26	5.05	5.45	5.35	5.16	5.58	5.44	5.28
95	TC	38.4	35.6	32.4	38.4	36.0	33.1	38.6	36.3	33.8
	SHC	20.1	26.3	31.2	20.4	27.6	32.5	21.2	28.7	33.5
	KW	5.71	5.49	5.25	5.79	5.60	5.38	5.87	5.69	5.50
100	TC	37.3	34.5	31.3	37.3	34.8	32.1	37.5	35.1	32.7
	SHC	19.7	25.9	30.4	20.1	27.0	31.7	21.0	28.4	32.5
	KW	5.84	5.59	5.35	5.91	5.70	5.48	6.00	5.80	5.61
105	TC	36.2	33.2	30.2	36.3	33.6	31.0	36.4	33.9	31.7
	SHC	19.4	25.3	29.7	20.0	26.7	30.8	20.5	28.0	31.6
	KW	5.97	5.69	5.44	6.04	5.80	5.58	6.13	5.90	5.72
115	TC	34.1	30.8	28.1	34.1	31.2	28.9	34.2	31.4	29.6
	SHC	18.6	24.5	28.1	19.3	25.9	28.9	19.9	27.2	29.6
	KW	6.21	5.89	5.63	6.28	6.00	5.78	6.38	6.10	5.94

Figure 4.1
Manufacturers Cooling Performance Data

than enough sensible and latent capacity. However, note that the capacities for Unit 6 match the loads even closer. Unit 6 must be operated above its nominal CFM to obtain the desired capacities. Unit 6 is smaller and it will provide the desired cooling at a lower first cost.

Step 3 - Determine the thermal balance point.

NOTE: This step is not necessary for designs where the cooling load is higher than the heating load.

The balance point is determined by constructing a diagram which correlates the building load and the unit capacity with the outdoor temperature. This diagram can always be constructed from the load calculation information and from the manufacturers performance data. The procedure is fairly simple for residential structures because the building load line is easy to draw. Conversely, the commercial load temperature diagram is not always simple to construct. (Refer to Section 1.6 for more information regarding this subject.) The main points to consider are summarized below.

Residential system considerations:

- Residential systems are normally not subject to large internal loads, they seldom include provisions for mechanical ventilation, and they usually do not have control systems that switch the equipment operating modes in accordance with changes of occupancy. Under these circumstances, the load line construction method discussed in Section 1.6 is fairly accurate.
- Always use the integrated capacity when plotting the equipment performance line.
- A unit can be oversized for cooling by as much as 25 percent if the extra capacity is required to obtain a lower heating balance point.

Example: Balance point for a residential system with no ventilation.

Residential Design Load = 40,000 Btuh @ 0°F OAT

Ventilation (Outdoor Air) CFM = 0

Check the balance point for the integrated heating capacities for Units 6 and 7, which were previously selected from the cooling requirements.

Figure 4.2 shows that unit 6 will have a balance point of about 28°F and the balance point for unit 7 is only slightly lower. Unit 6 is a good selection because it meets both the heating and cooling requirements. (Refer to Section 1.5 through 1.7)

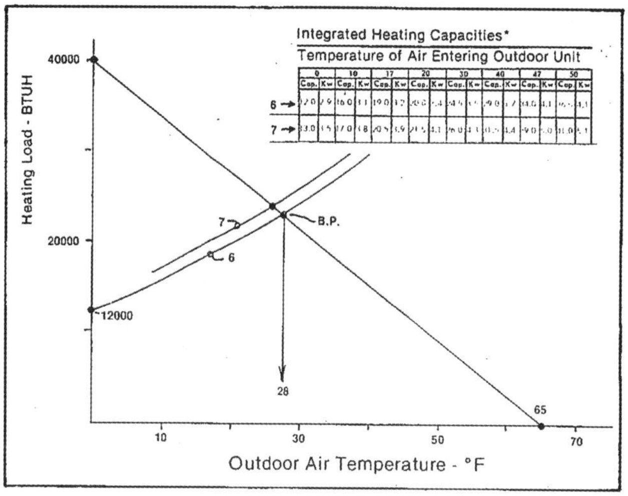

Figure 4.2 Balance Point Diagram, Residential

Commercial System Considerations:

- The load diagram for a commercial building cannot always be approximated by simply drawing a line from the design point to the 65°F "no load" point. In commercial work, the heating load diagram is affected by ventilation, occupancy, and internal loads. (Refer to Section 1.6)
- Since the heating system must be designed for the largest possible heating load, the designer is usually obligated to check two conditions:

  1) The occupied condition, which includes the occupied ventilation load, credit for the heat added by the internal loads and no credit for solar loads.

  2) The unoccupied condition, which normally does not include credit for internal and/or solar loads, but should include damper leakage loads. (From the standpoint of energy conservation, the outdoor damper should be closed when the building is unoccupied unless process requirements or codes specify otherwise. If ventilation is required when the building is unoccupied, the ventilation load must be included.)

Example: Balance point for a Commercial System or any system that has a provision for ventilation.

Heating Load = 40,000 Btuh @ 0°F (OAT)
Room Temperature = 70°F

Unit CFM = 1,300 & Ventilation CFM = 200

Damper Leakage = 5% (Closed)

Internal Heat Gain = 8,000 BTUH

Ventilation Load = 1.1 X 200 X (70 - 0) = 15,400 Btuh

Leakage Load = 1.1 X (.05 X 1,300) X (70 - 0) = 5,000 Btuh

Occupied Design Load = 40,000 + 15,400 -8,000 = 47,400 Btuh

Unoccupied Design Load = 40,000 + 5,000 = 45,000 Btuh

Figure 4.3 shows that Unit 6 will have a balance point of about 30°F for either the occupied or unoccupied condition and the balance point for Unit 7 is only slightly lower. (Note that if full ventilation is maintained when the building is vacant, the design balance point will rise to about 34°F.)

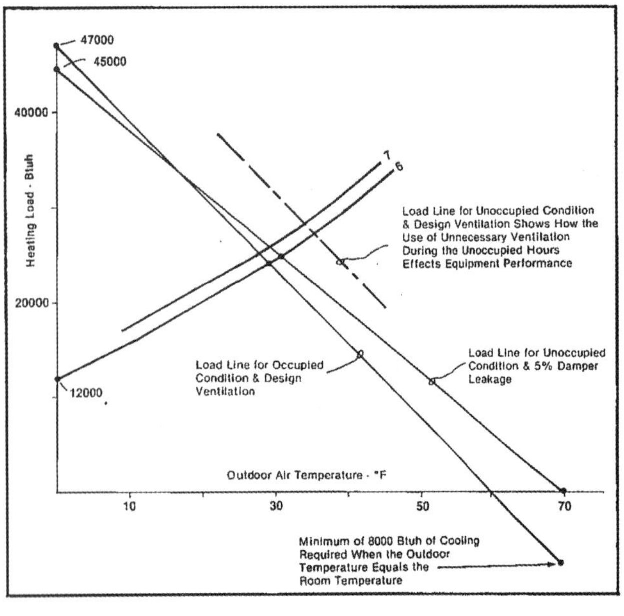

Figure 4.3 Balance Point Diagram, Commercial

The balance point for this example happened to fall near 30°F. Equipment that serves commercial buildings that have relatively high internal loads may have a design balance point that is well below 30°F. (An extreme example would be an internal zone that has little or no heating load. In this case the balance point could be well below zero °F.) No problems should occur as long as the unit is properly sized for cooling.

Step 4 - Determine the auxiliary heat requirement.

The auxiliary heat is the difference between the building load and the output of the heat pump when the outdoor temperature equals the heating design temperature. Refer to Figures 4.2 and 4.3 to see that the auxiliary heat required for the residential example is (40,000 - 12,000) = 28,000 Btuh, and the auxiliary heat required for the commercial example is (47,000 - 12,000) = 35,000 Btuh.

Figure 4.4, and the information shown below, illustrate how the supplemental heating coils are selected from the manufacturers data. The data shows that coil "C" is the best choice. Note the minimum CFM that is required to keep the coil from overheating. Also note that the coil comes in a single stage or two stage configuration. Refer to Section 3.10 for a discussion of the advantages of staging the auxiliary heaters.

Required Auxiliary Heat = 28,000 Btuh

Heating KW = 28,000 / 3413 = 8.2 KW
(3413 BTUH = 1 KW)

208 V supply

Manufacturers Supplemental Heater Data

	FAN COIL WITH ELECTRIC HEATER (208/240 v – 1 Ph)										
	Kw		Circuits			MWA		Capacity (1000 Btuh)			
			No.	Amps (ea)							
Coil	240 v	208 v		240 v	208 v	240 v	208 v	240 v	208v	Min Fan Speed*	Min Cfm
A	7.5	5.60	1	33.6	29.6	42.0	37.0	25.6	19.2	Med	
B	10.0	7.50	1	43.6	38.3	54.4	47.9	34.1	25.6	Med	
C	12.0	9.00	1	51.6	45.1	64.5	56.4	41.0	30.8	Med	
			2	20.0	17.3	25.0	21.6				
				28.0	26.0	35.0	32.5				1000
D	15.0	11.25	1	63.6	55.6	79.5	69.5	51.2	38.5	Med	
			2	20.0	17.3	25.0	21.6				
				43.6	38.6	54.5	48.2				
E	20.0	15.00	1	83.6	72.9	104.5	91.0	68.3	51.2	Med	
			2	40.0	35.0	50.0	43.7				
				43.6	38.6	54.5	48.2				

Residential Blower Performance Data for Unit (6)

CFM	Fan Speed		
	High	Medium	Low
	External Static Pressure (in. wg)*		
900	—	.80	.54
1000	.85	.72	.19
1100	.78	.63	—
1200	.68	.49	—
1300	.57	.27	—
1400	.43	—	—
1500	.27	—	—

Figure 4.4 Manufacturers Data - Supplemental Heat

Fan Performance

CFM	External Static Pressure (in. wg)															
	0.40		0.60		0.80		1.00		1.20		1.40		1.60		1.80	
	Rpm	Bhp	Rpm	Bhp	Rpm	Bhp	Rpm	Bhp	Rpm	Bhp	Rpm	Bhp	Rpm	Bhp	Rpm	Bhp
7,000	913	3.39	960	3.76	1012	4.20	1058	4.60	1105	5.34	1150	5.50	1195	6.00	1241	6.55
8,000	1012	4.94	1060	5.24	1101	5.66	1147	6.16	1187	6.62	1229	7.16	1269	8.06	—	—
9,000	1110	6.80	1160	7.05	1199	7.55	1236	8.02	—	—	—	—	—	—	—	—
9,000	885	4.76	940	5.48	992	6.19	1025	6.65	1098	7.82	1142	9.19	1187	9.09	1231	9.55
10,000	964	6.21	1009	6.94	1060	7.77	1104	8.52	1133	9.01	1196	10.08	—	—	—	—
11,000	1046	7.98	1082	8.70	1127	9.57	1173	10.50	1213	11.31	—	—	—	—	—	—
11,000	879	5.94	924	6.52	969	7.10	1013	7.70	1056	8.35	1097	9.00	1140	9.76	1178	10.48
12,000	943	7.50	985	8.10	1026	8.73	1067	9.38	1107	10.03	1147	10.75	1184	11.44	1223	12.22
13,000	1008	9.31	1047	9.96	1085	10.63	1123	11.31	1161	12.00	1198	12.72	1235	13.51	1269	14.25
14,000	1078	11.41	1140	12.10	1146	12.77	1182	13.53	1217	14.27	1251	15.03	1285	15.70	—	—
15,000	1141	13.82	1178	14.55	1202	15.23	1241	15.95	1275	16.80						

Figure 4.5 Blower Performance Data

Step 5 - Determine the fan performance

On residential work, the fan performance must be determined before the Manual D method can be used to size the duct system. Figure 4.5 shows the fan performance of the air handler used with Unit 6. Note that at 1300 CFM, the external static pressure is only 0.57 In. W.G. This is all the pressure that is available to move the air from the unit to the room and back to the unit.

On commercial work, the duct system should be sized by the equal friction method before the fan is selected. Commercial equipment comes in a wide range of fan performance options. Figure 4.6 illustrates the fan performance data for a commercial heat pump.

Indoor Air Fan Pulley Data

Fan RPM	Motor Pulley	Fan Pulley	Belt No. - Size
	No. Groves - Type - In.		
925	2 - 3V - 5.6	2 - 3V - 10.6	2 - 3V - 710
1093	2 - 3V - 5.0	2 - 3V - 8.0	2 - 3V - 670
1300	2 - 3V - 6.0	2 - 3V - 8.0	2 - 3V - 710
919	2 - 3V - 5.6	2 - 3V - 10.6	2 - 3V - 710
1088	2 - 3V - 5.0	2 - 3V - 8.0	2 - 3V - 670
1305	2 - 3V - 6.0	2 - 3V - 8.0	2 - 3V - 670
925	2 - 3V - 5.6	2 - 3V - 10.6	2 - 3V - 750
1093	2 - 3V - 5.0	2 - 3V - 8.0	2 - 3V - 670
1300	2 - 3V - 6.0	2 - 3V - 8.0	2 - 3V - 710

Figure 4.6 Pulley Data

Step 6 - Select the accessories and controls.

Accessories and controls include thermostat packages, disconnects, special relays, cycle timers, liquid line filter/dryers, trouble lights, mounting bases, return air plenums, refrigerant tubing packages, condensate pans, special mounting kits, various safety controls, and more. Each manufacturers catalog lists the options available.

4.6 Equipment Selection, Water to Air Equipment

The procedures described in Section 4.5 for air to air equipment also apply to water to air equipment. However, the entering temperature of the source water must be determined for both the heating and cooling season before the manufacturers data can be used to size the equipment. The water pressure drop through the water side heat exchanger will be an additional design consideration. Figure 4.7 illustrates an example of the manufacturers data for water to air equipment.

Cooling and Heating Capacities

Mode	CFM	FAN WATTS/ BF	EAT. (F)		208/230, 460 Volt								
					Ewt (F)								
					60			75			90		
					Gpm								
					7.5	10.5	14	7.5	10.5	14	7.5	10.5	14
Cooling	1300	690/0.10	72 WB	TC	46.3	43.7	41.8	45.6	44.3	43.2	43.2	43.3	43.0
				SHC	22.4	21.1	20.1	22.5	21.7	21.1	21.7	21.6	21.4
				KW	4.43	4.21	4.07	4.89	4.68	4.56	5.31	5.16	5.05
				THR	60.2	57.1	54.6	61.1	59.1	57.6	59.9	59.6	58.9
			67 WB	TC	44.3	42.3	40.3	42.2	42.0	41.0	37.5	39.7	40.1
				SHC	29.8	28.2	26.6	29.3	28.9	27.9	27.5	28.4	28.5
				KW	4.37	4.15	4.01	4.74	4.59	4.46	5.13	5.00	4.93
				THR	58.1	55.5	53.0	57.2	56.5	55.1	53.6	55.5	55.7
			62 WB	TC	41.2	40.5	39.2	37.9	39.1	38.9	32.3	34.8	36.2
				SHC	36.7	35.7	34.2	35.2	35.8	35.3	32.0	33.5	34.3
				KW	4.25	4.09	3.97	4.60	4.49	4.39	5.01	4.87	4.79
				THR	54.6	53.4	51.7	52.5	53.3	52.8	48.1	50.2	51.3
Heating	1300	690/0.10	70 DB	TC	37.9	39.5	40.5	46.3	46.6	47.4	50.8	51.8	52.4
				KW	4.55	4.62	4.64	4.96	4.98	5.02	5.24	5.30	5.33
				THA	26.3	27.8	28.8	33.9	34.2	34.8	37.8	38.6	39.1

BF — Bypass Factor
EAT — Entering Air Temperature
Ewt — Entering Water Temperature
Gpm — Gallons per minute
KW — Total power input (Compressor and Fan)
SHC — Sensible Heat Capacity (1000 Btuh)
TC — Total Net Capacity (1000 Btuh)
THA — Total Heat of Absorption
THR — Total Heat Rejection (1000 Btuh)

Heat Exchanger Pressure Drop
Water Side Pressure Drop VS Flow Rate
50HQ,VQ014-080

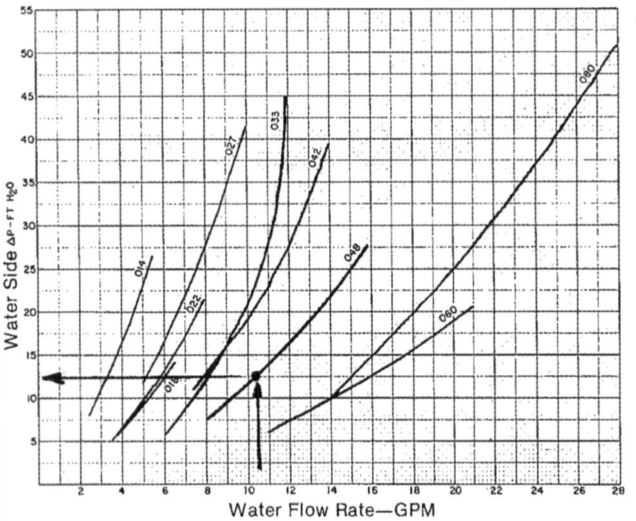

Figure 4.7 Water to Air Manufacturers Data

4.7 Equipment Selection, Air to Water or Water to Water Equipment

These heat pumps supply warm water or cold water to secondary heating and cooling devices. These devices may be located in the occupied space (fan coils, unit ventilators, unit heaters, convectors or induction terminals), or remotely located (central air handler). These heat pumps can also be used to supplement an existing boiler. application of this type of heat pump equipment requires knowledge of hydronic and chilled water design principles.

Sizing Secondary Air Side Equipment

The air side devices must provide adequate sensible and latent cooling capacity, proper air distribution and adequate heating capacity. The selection procedure is similar to the procedure outlined in Section 4.5 except the water coils provide the cooling or heating. Each air side terminal device (fan coil, air handler, unit ventilator etc.) will have heating and cooling capacities cataloged for various water flow rates and water temperatures. Supplemental heat may, or may not, be included in the air side equipment. The air side device may, or may not, include a duct system. However, in every case, the device must provide adequate air distribution and ventilation.

Sizing The Heat Pump

The heat pump capacities must match the peak heating and cooling loads that are associated with the secondary air side equipment. If one heat pump is used to serve terminal units located in more than one zone, the sizing calculation can take advantage of any diversity in the zone loads. In this case, a block load can be used to determine the heat pump capacity. If diversity is not applicable, the heat pump capacity must be based on the sum of the zone loads.

If the cooling loads are larger than the heating loads, the unit size will be determined by the cooling requirement. If the heating load is larger than the cooling load, a unit that is sized for heating will be oversized for cooling, and it will tend to short cycle on the water temperature or low pressure safety controls. Therefore, the unit cooling capacity should not exceed the design cooling load by more than 25 percent.

Water to water, or air to water heat pumps that are used to supplement a hot water boiler are not usually operated below the thermal or economic balance point. In this case, the heat pump capacity is determined by the heating load that occurs at the changeover condition. Since the secondary heating devices were originally sized for higher boiler water temperatures (150 to 200°F), they will produce less heat at the heat pump water temperatures (85 to 130°F). The thermal balance point for add on equipment is usually determined by the heating capacity of the existing terminal devices when they operate at the water temperatures produced by the heat pump.

Auxiliary Heat

The auxiliary heat should compensate for any difference between the space load and the output of the heat pump at the design heating load condition. Supplementary heat can be supplied by adding heat at the secondary heating device, or directly to the supply water. (Adding the heat at the terminal device allows a zoned system to take advantage of diversity in the heating loads.)

4.8 Heat Pump Energy and Operating Cost Calculations

Heat pump energy calculations are usually made by using some form of the bin method. The degree day method should not be used to estimate heat pump energy requirements.

When the bin method is used for commercial, energy estimates, the procedure usually requires separate bin calculations for each operating condition. Examples of some possible operating conditions include:

Mechanical Cooling (Occupied). Minimum ventilation CFM, credit for internal gains and effects of solar loads.

Mechanical Cooling (Unoccupied). Outdoor air damper closed, credit for higher thermostat setpoint and effects of any solar loads.

No Cooling (Unoccupied). Includes effects of thermal storage during next cooling cycle.

Free Cooling (Occupied). Economizer cycle with dry bulb or enthalpy control.

Heating (Occupied). Minimum ventilation CFM, credit for internal gains and the effects of solar loads.

Heating (Unoccupied). Outdoor dampers closed, credit for lower thermostat setpoint and effects of any solar loads.

No Heating (Unoccupied). Includes the effects of thermal storage during the next heating cycle.

Operating conditions are also affected by other factors such as capacity control, hot gas bypass, reheat, setpoint schedules, ventilation schedules, demand control, etc.

A discussion of energy calculations for commercial comfort or reclaim heat pump applications is beyond the scope of this manual. (The bin calculation procedure requires a determination of the bin energy requirement for each possible operating condition and a determination of the actual number of bin hours per year that a given condition occurs.)

The bin method can be readily applied to residential heat pumps because internal loads are small and relatively constant. Ventilation and zoning are usually not a factor because the equipment operates in the same mode during any 24 hour period. The methodology for residential bin calculations is discussed below.

4.9 Residential Air Source Heat Pump Energy Calculation

This method will allow for variations in the heat pump COP as the source temperature changes and for any energy required for defrost and auxiliary heat.

It is assumed that no heat is required when the outdoor temperature is 65°F or higher and that the full design load occurs at the winter design temperature. In between these two extremes, the heating load is assumed to vary linearly with the outdoor air temperature.

The heating energy required for a given temperature bin is equal to the heat pump power plus any auxiliary heating power that is required for that bin,multiplied by the number of hours per year that the bin temperature occurs.

Example:

A residence has 1,007 hours per year when the outdoor temperature falls between 40°F and 50°F. When the outdoor temperature is at the midpoint of the Bin (45°F), the heat pump requires 4.1 KW and runs only 49 percent of the time.

Bin Energy = 1,007 Hrs/Yr X 4.1 KW X 0.49 = 2,023 KWH/YR

This calculation can be made for each bin that occurs between the winter design temperature and 65°F. The total energy required equals the sum of the energy consumed in each individual bin. The

Figure 4.8 and the instructions on page 57 illustrate how the bin calculation is performed.

Figure 4.8 Residential Air Source Energy Calculation

Step 1 — Draw the Heating Load vs. OAT Diagram and Integrated Capacity vs. OAT Diagram.

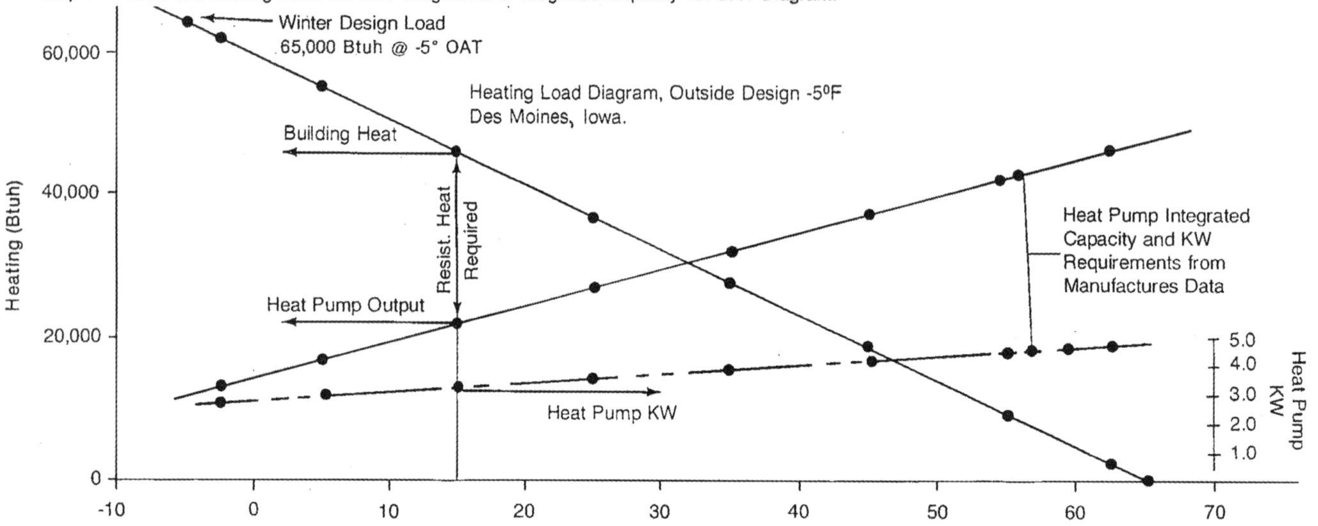

Heat Pump Operates Below Balance Point

Outside Air Temperature (°F)

Case I — Temperature Bins →	-5 to 0	0 to 10	10 to 20	20 to 30	30 to 40	40 to 50	50 to 60	60 to 65	Totals	Units
Line 1 — Bin Weather Data	56	212	475	992	1,393	1,007	1,118	694		Hr./Yr
Line 2 — Building Heat Required ÷ 1,000	62	55	47	37	28	18	10	2		Btuh
Line 3 — Heat Pump Capacity ÷ 1,000	13	17	22	27	32	37	42	50		Btuh
Line 4 — Resistance Heat Required ÷ 1,000	49	38	25	10	0	0	0	0		Btuh
Line 5 — Heat Pump Power	2.8	3.0	3.3	3.6	3.8	4.1	4.4	4.7		KW
Line 6 — % Time Heat Pump Runs	1.0	1.0	1.0	1.0	.875	.486	.238	.040		—
Line 7 — Heat Pump Energy ÷ 1,000	157	636	1,567	3,571	4,631	2,006	1,171	130	13,869	Kwh/Yr.
Line 8 — Resistance Heating Power	14.4	11.1	7.3	2.9	0	0	0	0		KW
Line 9 — Resistance Heating Energy ÷ 1,000	806	2,353	3,468	2,877	0	0	0	0	9,504	Kwh/Yr.
Total Energy Required = Line 7 + Line 9 =									23,373	Kwh/Yr.

Heat Pump Off Below the Balance Point

Outside Air Temperature (°F)

Case II — Temperature Bins →	-5 to 0	0 to 10	10 to 20	20 to 30	30 to 40	40 to 50	50 to 60	60 to 65	Totals	Units
Line 1 — Bin Weather Data	56	212	475	992	1,393	1,007	1,118	694		Hr./Yr
Line 2 — Building Heat Required ÷ 1,000	62	55	47	37	28	18	10	2		Btuh
Line 3 — Heat Pump Capacity ÷ 1,000	—	—	—	—	32	37	42	50		Btuh
Line 4 — Resistance Heat Required ÷ 1,000	62	55	47	37	0	0	0	0		Btuh
Line 5 — Heat Pump Power	OFF	OFF	OFF	OFF	3.8	4.1	4.4	4.7		KW
Line 6 — % Time Heat Pump Runs	—	—	—	—	.875	.486	.238	.040		—
Line 7 — Heat Pump Energy ÷ 1,000	—	—	—	—	4,631	2,006	1,171	130	7938	Kwh/Yr.
Line 8 — Resistance Heating Power	18.2	16.2	13.8	10.9	0	0	0	0		KW
Line 9 — Resistance Heating Energy ÷ 1,000	1019	3434	6555	10813	0	0	0	0	21821	Kwh/Yr.
Total Energy Required = Line 7 + Line 9 = TOTAL =									29759	Kwh/Yr.

Residential Air Source Bin Calculation Procedure

(Section 4.9 Continued)

Draw the heating load/temperature line.

Draw the heat pump integrated capacity/temperature line.

Draw the heat pump power (KW)/temperature line.

Line 1 - enter hours per year for each bin.

Line 2 - enter the heating load for each bin.

Line 3 - enter heat pump capacity for each bin.

Line 4 - subtract line 3 (heat pump capacity) from line 2 (heating load) to find the amount of auxiliary heat required.

Line 5 - enter the heat pump KW requirement for each bin.

Line 6 - enter the heat pump running time for each bin. This is calculated by dividing line 2 by line 3. Use 1.0 for the bins that have loads that are greater than the heat pump capacity.

Line 7 - enter the heat pump KWH for each bin. This is calculated by multiplying lines 1, 5 and 6 together. Add the bin KWH values together to find the yearly heat pump energy requirement.

Line 8 - enter the resistance heat KW for each bin. This is calculated by dividing line 4 by 3.413.

Line 9 - enter the resistance heat KWH for each bin. This is calculated by multiplying lines 1 and 8 together. Add the bin KWH values together to find the yearly resistance heat energy requirement.

The total KWH required is the sum of the heat pump KWH and the resistance KWH. In the example illustrated in Figure 4.8, 13,869 + 9,504 = 23,373 KWH/Yr.

4.10 Residential Ground Water Heat Pump Energy Calculation

The bin calculation procedure discussed above for the air to air heat pump can also be used for ground water applications. Residential water source heat pumps may be designed for heating only with cooling by a separate water coil, or for refrigeration cycle heating and cooling. (Refer to Section 2.3 for more information.) In both cases, the equipment capacity line and the balance point must be determined before the energy calculation can be made. (Refer to Figure 2.5 for examples of ground water heat pump performance. Note that the water source heat pump capacity and power requirements depend on the temperature of the source water and are relatively constant and independent of outdoor temperature.)

The bin calculation must also include the water pump power requirements. Assume that the water pump operates for the same number of hours that the heat pump operates. The pump power draw will vary with system gpm and pressure drop. Consult a well driller for estimates of the pump power.

Example: "Heating only" heat pump sized for winter design load.

The procedure for the "Heating Only" heat pump is the same as the one used for the air source calculation described previously except that lines 8 and 9 are not required because auxiliary heat is not used. Figure 4.9 illustrates the procedure.

Example: "Heating and cooling" ground water heat pump (size limited by the summer cooling load).

In this case, the balance point requirements must be considered and the procedure is the same as the one used for the air source calculation described above. Figure 4.10 illustrates the procedure.

Residential Water Source BIN Calculation: Heat Pump Sized for Heating Design Load

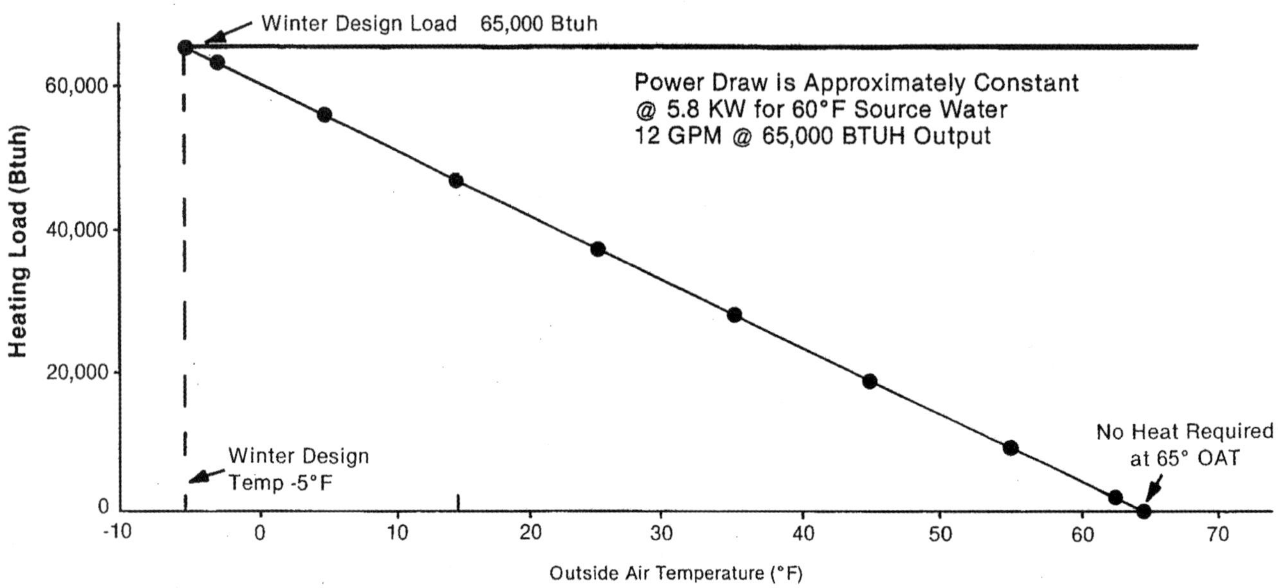

See Text for Procedure

Temperature Bins →	-5 to 0	0 to 10	10 to 20	20 to 30	30 to 40	40 to 50	50 to 60	60 to 65	Totals	Units
Line 1 — Bin Weather Data	56	212	475	992	1,393	1,007	1,118	694		Hr./Yr
Line 2 — Building Heat Required ÷ 1,000	62	55	47	37	28	18	10	2		Btuh
Line 3 — Heat Pump Capacity ÷ 1,000	65	65	65	65	65	65	65	65		Btuh
Line 4 — Resistance Heat Required ÷ 1,000	0	0	0	0	0	0	0	0		Btuh
Line 5 — Heat Pump Power	5.8	5.8	5.8	5.8	5.8	5.8	5.8	5.8		KW
Line 6 — % Time Heat Pump Runs	.95	.85	.72	.57	.43	.28	.15	.03		—
Line 7 — Heat Pump Energy ÷ 1,000	309	1045	1984	3279	3474	1635	648	121	12,495	Kwh/Yr.
Line 8 — Resistance Heating Power										KW
Line 9 — Resistance Heating Energy ÷ 1,000									9,504	Kwh/Yr.
									12,495	Kwh/Yr.
Pump Energy - 0.75 KW x line 1 x line 6	40	135	257	424	449	211	126	16	1658	Kwh/Yr.
								Total Kwh =	14,153	Kwh/Yr

Figure 4.9 Water Source Energy

Residential Water Source BIN Calculation: Heat Pump Sized With Balance Point

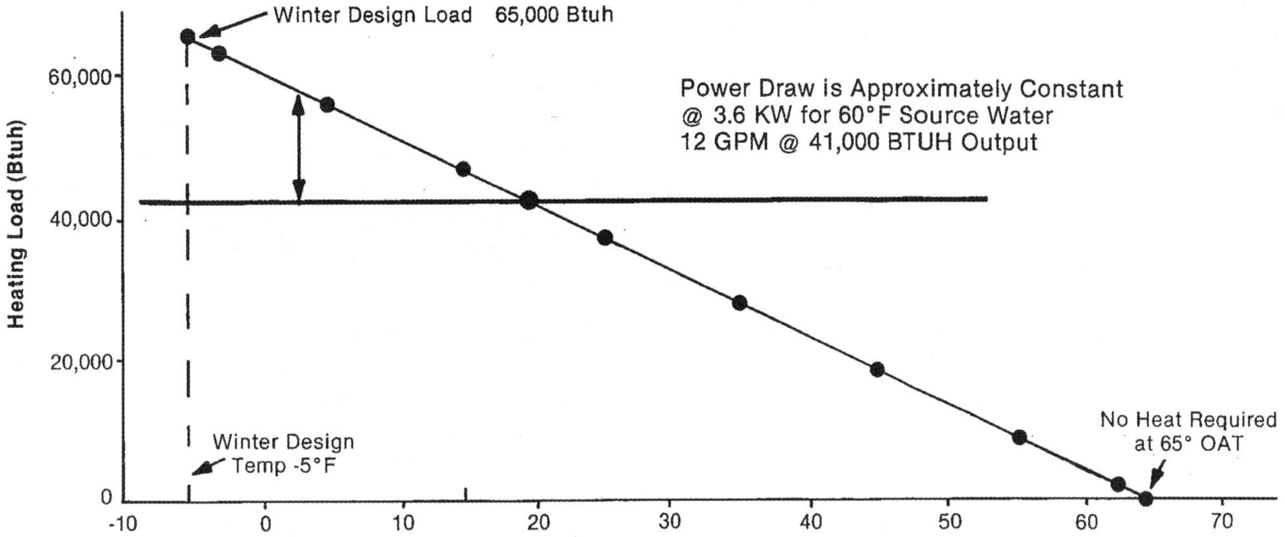

Outside Air Temperature (°F)

See Text for Procedure — Temperature Bins →	-5 to 0	0 to 10	10 to 20	20 to 30	30 to 40	40 to 50	50 to 60	60 to 65	Totals	Units
Line 1 — Bin Weather Data	56	212	475	992	1,393	1,007	1,118	694		Hr./Yr
Line 2 — Building Heat Required ÷ 1,000	62	55	47	37	28	18	10	2		Btuh
Line 3 — Heat Pump Capacity ÷ 1,000	41	41	41	41	41	41	41	41		Btuh
Line 4 — Resistance Heat Required ÷ 1,000	21	14	6	—	—	—	—	—		Btuh
Line 5 — Heat Pump Power	3.6	3.6	3.6	3.6	3.6	3.6	3.6	3.6		KW
Line 6 — % Time Heat Pump Runs	1.0	1.0	1.0	.90	.68	.44	.24	.05		—
Line 7 — Heat Pump Energy ÷ 1,000	202	763	1710	3214	3410	1595	966	125	11,985	Kwh/Yr.
Line 8 — Resistance Heating Power	6.2	4.1	1.8	—	—	—	—	—		KW
Line 9 — Resistance Heating Energy ÷ 1,000	347	869	855	—	—	—	—	—	2071	Kwh/Yr.
									14,056	Kwh/Yr.
Pump Energy - 0.75 KW x line 1 x line 6	40	135	257	424	449	211	126	16	1658	Kwh/Yr
								Total Kwh=	15714	Kwh/Yr

Figure 4.10 Bin Cooling Energy

4.11 Add on Heat Pump Energy Calculations

The bin method can also be used to estimate the annual energy consumption for add on units. There are three possible cases:

Case 1 - The economic balance point is equal to, or above, the thermal balance point. In this case, the furnace is off when the outdoor temperature is above the economic balance point and the heat pump is off when the outdoor temperature is below the economic balance point.

Case 2 - The economic balance point is below the thermal balance point and the controls prevent the heat pump from operating when the outdoor temperature is below the thermal balance point. In this case, the furnace is off when the outdoor temperature is above the thermal balance point and the heat pump is off when the outdoor temperature is below the thermal balance point.

Case 3 - The economic balance point is below the thermal balance point and the controls allow the heat pump to operate when the outdoor temperature is between the economic balance point and the thermal balance point. In this case, the heat pump is allowed to operate when the outdoor temperature is below the thermal balance point and above the economic balance point. When the space temperature drops too far below the desired setpoint, the heat pump is de-energized and the furnace operates until the thermostat is satisfied. Then, the furnace is cycled off and the heat pump is cycled back on.

Examples: Add on heat pump.

Cases 1 and 2 - Figures 4.11A and 4.11B

Above the design balance point (economic or thermal), the procedure is the same one used for the air source calculation described in Section 4.9, except lines 4, 8 and 9 are not required because auxiliary heat is not used.

Below the design balance point, multiply lines 1 and 2 together to determine the fossil energy required for each bin. Add all the bins together to determine the total amount of fossil fuel heat required for the structure. To determine the amount of raw fossil energy that must be purchased, divide the fossil energy that is required for the structure by seasonal efficiency of the fossil fuel furnace.

Case 3 - Figure 4.11C

Above the thermal balance point, the procedure is the same one used for the air source calculation described in Section 4.9, except lines 4, 8 and 9 are not required because auxiliary heat is not used.

Below the economic balance point, multiply lines 1 and 2 together to determine the heating energy required for each bin, then add all the bins together to determine the total amount of fossil fuel heat required for the structure.

Between the economic balance point and the thermal balance point, estimate the percentage of time heat pump runs by using the equation found on Figure 4.11c. Use this percentage on line 8 and carry out the heat pump energy calculation for this temperature range. Also estimate the percentage of time the furnace operates in this temperature range and enter the value on line 10. Multiply line 10 by lines 1 and 2 to determine the amount of fossil energy required.

To determine the amount of raw fossil energy that must be purchased, divide the fossil energy that is required for the structure by seasonal efficiency of the fossil fuel furnace.

Residential "Add on" Heat Pump BIN Calculation:

Case I - Economic B.P. Equal to or Above Theoretical Thermal B.P.

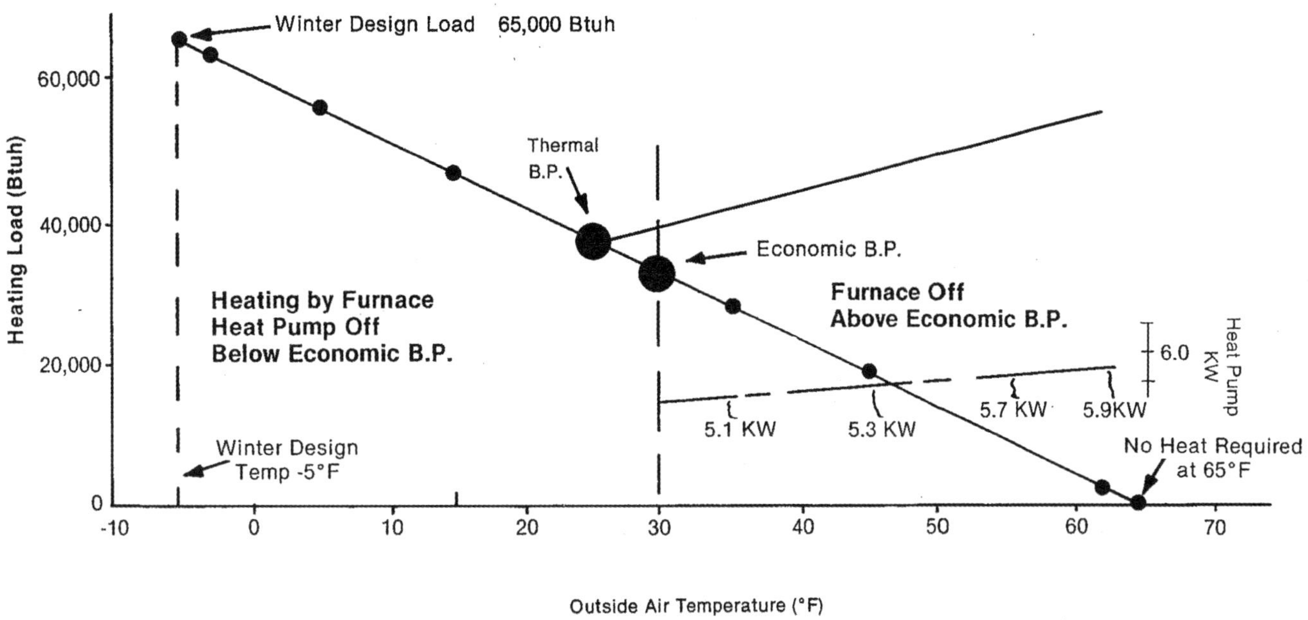

See Text for Procedure / Temperature Bins →	-5 to 0	0 to 10	10 to 20	20 to 30	30 to 40	40 to 50	50 to 60	60 to 65	Totals	Units
Line 1 — Bin Weather Data	56	212	475	992	1,393	1,007	1,118	694		Hr./Yr
Line 2 — Building Heat Required ÷ 1,000	62	55	47	37	28	18	10	2		Btuh
Line 3 — Heat Pump Capacity ÷ 1,000	—	—	—	—	42	47	52	55		Btuh
Line 4 — Resistance Heat Required ÷ 1,000	—	—	—	—	—	—	—	—		Btuh
Line 5 — Heat Pump Power	—	—	—	—	5.1	5.3	5.7	5.9		KW
Line 6 — % Time Heat Pump Runs	—	—	—	—	.67	.38	.19	.04		—
Line 7 — Heat Pump Energy ÷ 1,000	—	—	—	—	4760	2028	1211	164	8163	Kwh/Yr.
Line 8 — Resistance Heating Power										KW
Line 9 — Resistance Heating Energy ÷ 1,000										Kwh/Yr.

Fossel Fuel Energy Req'd	3472	11660	22325	36704

= 74,161,000 Btu/Yr

Line 1 x Line 2

If the Seasonal Furnace Efficiency is 0.65%
Raw Fossil Fuel Energy Required =
74,161,000/.65 = 114,094,000 BTU/YR

Heating Energy = 8163 KWH Electrical + 114,094,000 BTU Fossil

Figure 4.11-A

Residential "Add on" Heat Pump BIN Calculation:

Case II - Economic B.P. Below Thermal B.P. and Heat Pump Off Below Thermal B.P.

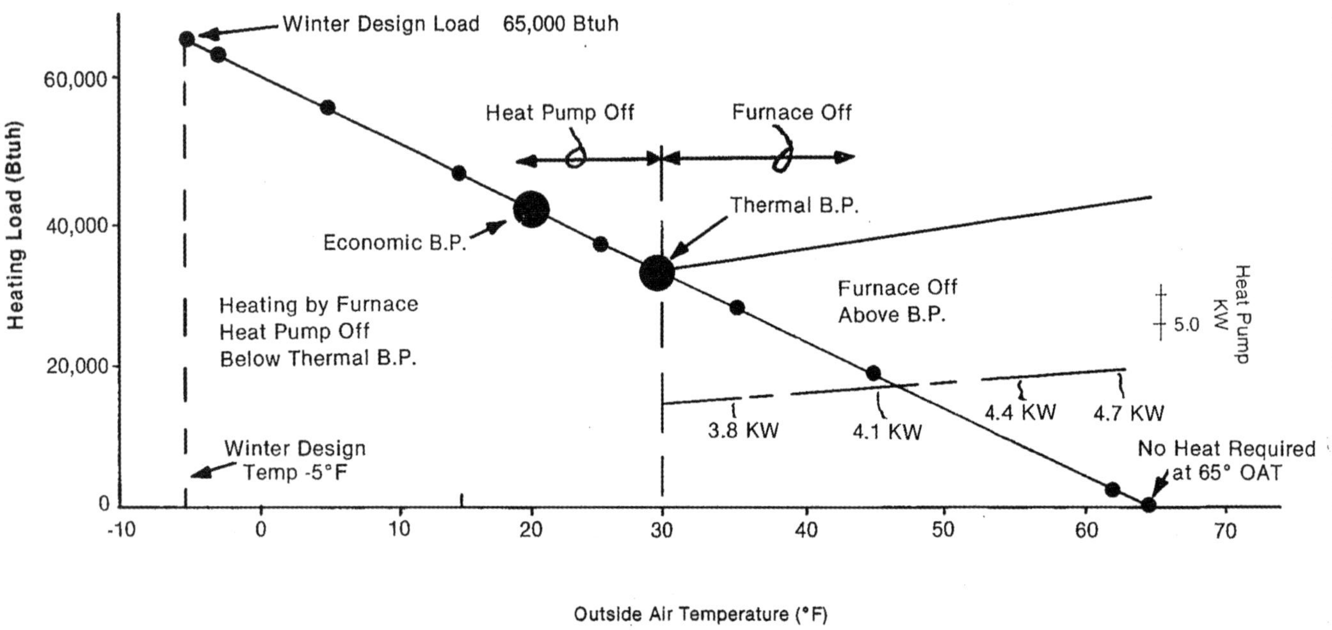

See Text for Procedure — Temperature Bins →	-5 to 0	0 to 10	10 to 20	20 to 30	30 to 40	40 to 50	50 to 60	60 to 65	Totals	Units
Line 1 — Bin Weather Data	56	212	475	992	1,393	1,007	1,118	694		Hr./Yr
Line 2 — Building Heat Required ÷ 1,000	62	55	47	37	28	18	10	2		Btuh
Line 3 — Heat Pump Capacity ÷ 1,000	—	—	—	—	32	37	42	50		Btuh
Line 4 — Resistance Heat Required ÷ 1,000	—	—	—	—	—	—	—	—		Btuh
Line 5 — Heat Pump Power	—	—	—	—	3.8	4.1	4.4	4.7		KW
Line 6 — % Time Heat Pump Runs	—	—	—	—	.875	.486	.238	.04		—
Line 7 — Heat Pump Energy ÷ 1,000	—	—	—	—	4631	2006	1171	130	7938	Kwh/Yr.
Line 8 — Resistance Heating Power										KW
Line 9 — Resistance Heating Energy ÷ 1,000										Kwh/Yr.

Fossel Fuel Energy Req'd	3472	11660	22325	36704

= 74,161 (1000) Btu/Yr for Structure

Line 1 x Line 2

If the Seasonal Furnace Efficiency is 0.65%
Raw Fossil Fuel Energy Required =
74,161,000/.65 = 114,094,000 BTU/YR

Heating Energy = 7938 KWH Electrical + 114,094,000 BTU Fossil

Figure 4.11-B

Case III - Economic B.P. Below Thermal B.P. and Heat Pump Operates When the Furnace is Off and the OAT is Below the Thermal B.P. but Above the Economic B.P.

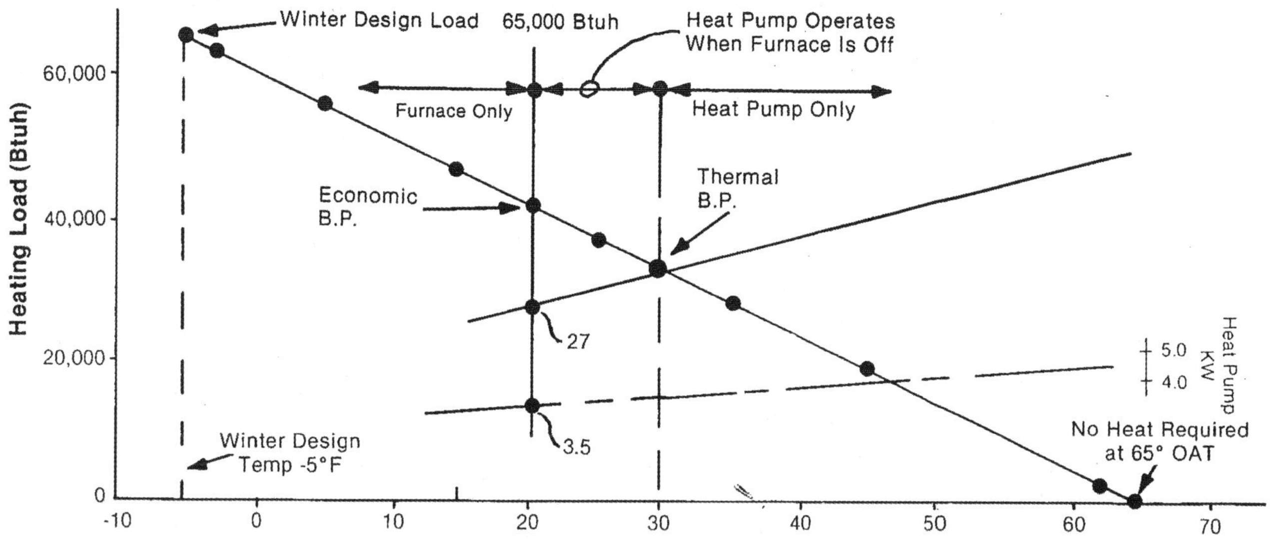

Outside Air Temperature (°F)

See Text for Procedure / Temperature Bins →	-5 to 0	0 to 10	10 to 20	20* to 30	30 to 40	40 to 50	50 to 60	60 to 65	Totals	Units
Line 1 — Bin Weather Data	56	212	475	992	1,393	1,007	1,118	694		Hr./Yr
Line 2 — Building Heat Required ÷ 1,000	62	55	47	37	28	18	10	2		Btuh
Line 3 — Heat Pump Capacity ÷ 1,000	—	—	—	27	32	37	42	50		Btuh
Line 4 — Resistance Heat Required ÷ 1,000	—	—	—	—	—	—	—	—		Btuh
Line 5 — Heat Pump Power	—	—	—	3.5	3.8	4.1	4.4	4.7		KW
Line 6 — % Time Heat Pump Runs	—	—	—	.74	.875	.486	.238	04		—
Line 7 — Heat Pump Energy ÷ 1,000	—	—	—	2569	4631	2006	1171	130	10507	Kwh/Yr.
Line 8 — Resistance Heating Power										KW
Line 9 — Resistance Heating Energy ÷ 1,000										Kwh/Yr.
Line 10 - % Heat Supplied by Furnace	1.0	1.0	1.0	.26	0	0	0	0		

$$* \ \% \text{ Time H.P. Runs} = \frac{\text{Winter Design Load - BIN Load}}{\text{Winter Design Load - BIN H.P. Capacity}} = \frac{65 - 37}{65 - 27} = 0.74 = 74\%$$

% Heat from furnace = 1 - % times H.P. runs = 1.0 - 0.74 = 0.26 = 26%

Fossel Fuel Energy Req'd	3472	11660	22325	9543

47,000,000 Btu/Yr for structure

Line 1 x Line 2 = Line 10

If the Seasonal Furnace Efficiency is 0.65%:
Raw Fossil Fuel Energy Required =
47,000,000 = 72,308,000 Btu/Yr

Heating Energy = 10,507 KWH Electrical + 72,308,000 Btu Fossil

Figure 4.11-C

4.12 Residential Cooling Energy Requirements

Cooling is usually accomplished mechanically with a DX coil, but some ground water heat pumps employ water coil cooling. The bin procedure can be applied to both cases. Refer to Figure 4.12.

Example: - Residential bin calculation procedure, DX cooling.

Draw the heating load/outdoor temperature line.

Line 1 - enter the bin cooling hours.

Line 2 - enter the bin cooling load.

Line 3 - enter the bin energy (Line 1 & Line 2).

Total all the bin energy to find the cooling energy required for the structure (21,397,000 Btu/Yr).

Calculate the amount of cooling energy that must be supplied by the unit by dividing the energy required for the structure by 1000 and by the SEER rating of the heat pump.

Example - Residential water coil cooling energy.

In this case, the water pump and the fan are the only devices that consume energy, and they will only operate for a percentage of the hours contained in any bin. This percentage can be approximated by dividing the cooling load for any bin by the design capacity of the unit. Estimate the total cooling energy by multiplying the pump and fan power draw (KW) by the total number of operating hours. (Refer to the cooling load line shown in Figure 4.12)

Fan and Pump Draw = 0.9 KW

Design Sensible Capacity = 30,000 Btuh

Bin		Bin Hrs	% On Time		Operating Hours
(70 To 75)	=	709	4/30	=	95
(75 To 80)	=	562	11/30	=	206
(80 To 85)	=	378	18/30	=	227
(85 To 91)	=	223	25/30	=	186

Total Operating Hours = 714

Total Cooling Energy = 0.9 KW * 714 Hours/Yr = 643 KWH/Yr

Residential BIN Calculation for DX Cooling

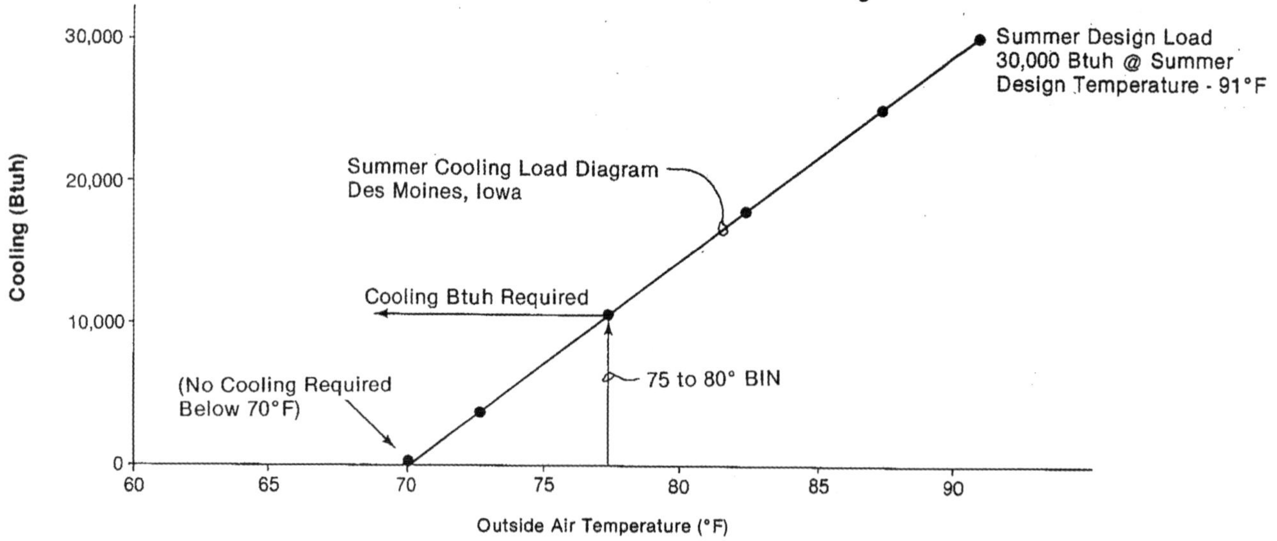

Procedure	Temperature Bins →	70 to 75	75 to 80	80 to 85	85 to 91	Units
Line 1 — Enter Bin Weather Data		709	562	378	223	Hrs./Yr.
Line 2 — Enter the Bin Cooling Load		4	11	18	25	1,000 Btuh
Line 3 — Line 1 × Line 2 = Energy per yr.		2,836	6,182	6,804	5,575	1,000 Btu/Yr.
Sum up the numbers on Line 3 and Multiply by 1,000 to find the Yearly Btu.		21,397,000				Btu/Yr.
Calculate Electrical Energy Required Cooling $\frac{Kwh}{Yr} = \frac{Btu/Yr.}{1,000 \times Seer}$		For Seer = 9.5 $\frac{Kwh}{Yr} = \frac{21,397,000}{1000 \times 9.5} = \frac{2,252\ Kwh}{Yr.}$				

Figure 4.12

Section 5

General Installation Procedures

The procedures described below are relatively general and apply to both residential and commercial equipment. This section only identifies the various installation considerations and is not intended to be used as an installation manual. Specific installation instructions for a particular piece of equipment should always be obtained from the equipment manufacturer.

In addition to the manufacturers recommendations, all equipment should be installed in accordance with codes, standards, ordinances and regulations which may include:

- National, state or local building codes.
- National Fire Protection Association standards (NFPA).
- Air Conditioning and Refrigeration Institute (ARI) standards.
- National, state or local electric codes.
- Local and/or state energy codes.
- Local utility standards and regulations.
- Local and/or state ordinances governing airborn noise and property setback requirements.
- Local and/or state ordinances which govern the use of city water and sewer systems.
- Local and/or state regulations which govern the use of aquifers, well development, and water disposal.
- National, state, or local plumbing codes.
- Standards, codes or regulations that apply specifically to government buildings.

All the pertinent codes, regulations, standards and ordinances should be collected in one notebook, including a summary of the main points of each document.

5.1 Outdoor Equipment Location, Clearances, and Support

Outdoor air source heat pump equipment may be located on the ground, on a frame, or on a roof. The following considerations apply:

- Air cooled condensers must have access to an unrestricted supply of outdoor air. Design air flow through the outdoor coil must be maintained under all operating conditions and the air must not recirculate through the outdoor coil.
- Avoid placing the outdoor unit in locations where it will be exposed to hot building exhaust air, or where heat is generated by other equipment, or in locations where pavement or masonry materials produce a high ambient temperature (Figure 5.1). (An exception is made when the equipment is specifically designed to reclaim waste heat but this involves much more than simply setting the unit in a warm environment.)

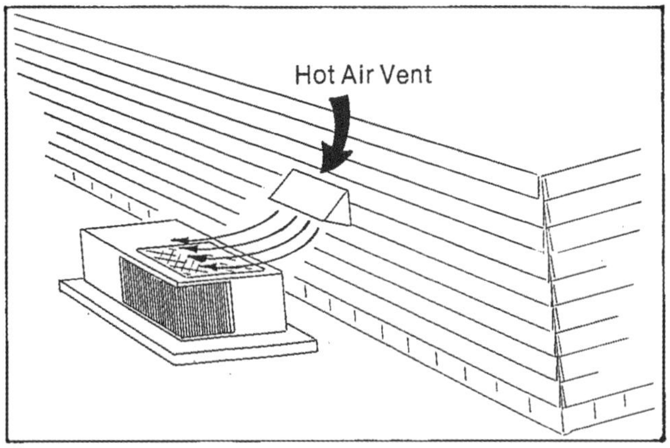

Figure 5.1 Avoid Hot Air Vents

- Consider the effects of noise on the occupants and the neighbors.
- Provide adequate clearances for making all connections, for servicing the unit, and for replacing the components contained in the unit.
- Locate the unit so that the refrigerant piping can be installed in accordance with the manufacturers recommendations.
- Consider the availability of the electrical supply.
- Protect the outdoor unit from winter winds to insure correct defrost system operation (Figure 5.2).

Figure 5.2 Protect From Wind

- Provide a foundation pad for units that are located at ground level (Figure 5.3). In cold climates, provide a pad and a suitable frame to elevate the unit above snow accumulations (Figure 5.4). Make sure that the building framing is adequate to support rooftop equipment and remember to elevate the unit above snow accumulation on the roof (Figure 5.5).

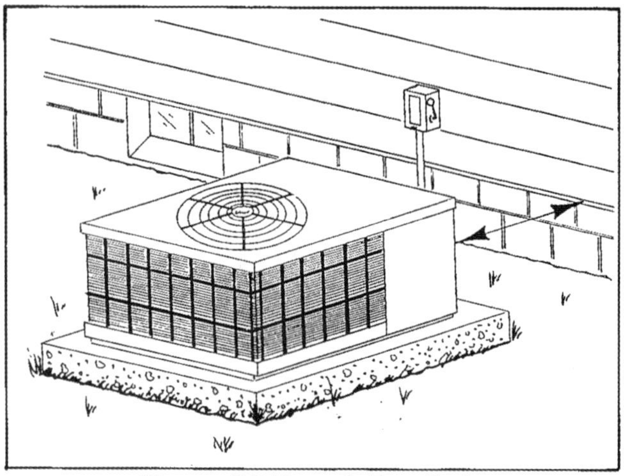

Figure 5.3 Foundation Pad

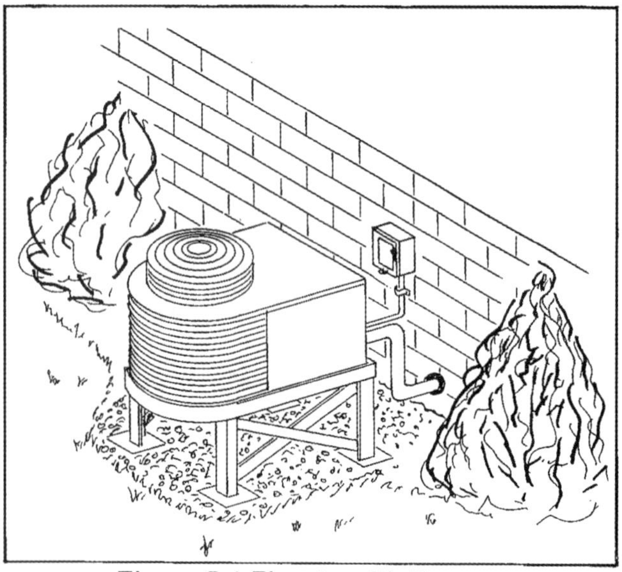

Figure 5.4 Elevate with Frame

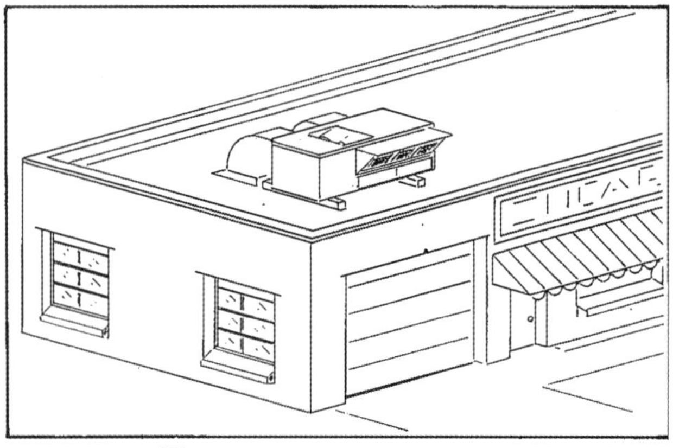

Figure 5.5 Use Roof Curbs

- Aesthetics may be important. Consider the architectural impact of the equipment location.

5.2 Indoor Equipment Location, Clearance, and Support

Indoor airside equipment includes the air handler section and, in some cases, a compressor section. Indoor air side equipment can be located in the basement, attic, equipment room or in a closet (Figure 5.6). Air handlers are available in both vertical and horizontal configurations. Packaged "water source" and "air to water" equipment is also usually located indoors.

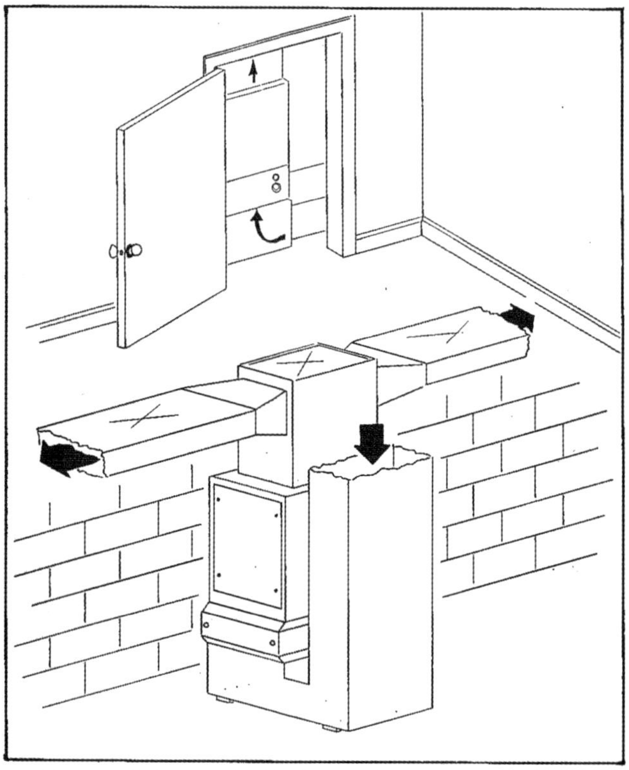

Figure 5.6 Locate Unit in Closet or Basement

General Comments

- Provide adequate clearances for making all connections, for servicing the unit and for replacing the components contained in the unit.
- Consider the availability of the electrical supply.
- Consider equipment noise which may be transmitted through walls, floors, or ceilings.
- Inertia pads and vibration isolators may be required.

Air Side Equipment Location

- Locate the unit so that the refrigerant piping can be installed in accordance with the manufacturers recommendations.

- Position the air handler so the duct system will not have unusually long runs or inefficient fittings.
- If ventilation is required, the unit must be located so that outdoor air and exhaust air can be ducted to and from the unit.
- Consider the availability of a condensate drain

Water to Water and Air to Water Equipment Location

- Water side equipment should not be exposed to freezing temperatures without some form of anti-freeze protection.
- Provide clearance so water side heat exchangers can be disassembled and cleaned.
- Water to water equipment location should be convenient to the water supply system and should not require unusually complicated piping runs or result in excessive pumping heads.
- Air to water equipment location should be convenient to the source of outdoor air that must be ducted to and from the unit.
- The unit should be located so that it can be easily connected to the hydronic system.

5.3 Air Distribution Systems

Duct systems should be installed in accordance with all applicable codes and/or the Sheet Metal and Air Conditioning Contractors National Association (SMACNA) standards. Fire dampers may be required in some cases. Outdoor duct systems must be insulated and protected against the weather. Vibration isolation and sound attenuation are additional considerations.

5.4 Water Piping Systems

Water piping systems should be installed in accordance with all applicable codes and standards. Well water systems should be installed in accordance with local and state regulations, manufacturers and/or the National Well Water Association recommendations. Other considerations include.

- Pump down tests to verify well capacities.
- Freeze protection.
- Pressure tank size, charge, location and detail piping.
- Make-up water and back flow prevention.
- Insulation to prevent condensation on indoor piping.
- Vibration isolation and flexible connections for pumps and piping.
- Manual or automatic control of water flow rates (water regulating valve to control head pressure on water source heat pumps).
- Isolation valves (allow equipment service without draining the system).
- Slow closing solenoid valve (to shut off water flow when the ground water heat pump compressor is off).
- Hose bibbs (to flush, drain or fill the system).
- Temperature and gauge ports (for testing, adjusting and balancing the water system).
- Water quality (the water system should not be subject to unusual biological growth, scale, corrosion, or suspended particles).
- Cupro-nickel heat exchangers (for water that has a high salt or mineral content, or if biological growth or scale is anticipated).
- Foreign materials (the system should be cleaned, flushed and free of debris before it is put into operation).
- Provisions for manually and automatically purging air from the system.
- PVC, and polethylene piping can be used outside the building on ground water and earth coil heat pump systems. (Piping must be below the frost line.)
- On ground water systems, either PVC, polybutylene, rubber and copper are recommended for connecting the pressure tank to the heat pump. (Steel and galvanized pipe is subject to corrosion and transmits noise.)

5.5 Refrigerant Piping

All refrigerant piping should be performed in accordance with the manufacturers recommendations. The manufacturer will specify the maximum recommended lengths for horizontal runs, or risers, and the corresponding minimum pipe or tubing diameters. Other refrigerant piping considerations include:

- **Contamination** - During installation the piping must be kept free of metal particles, dirt, moisture, carbon deposits, and other foreign materials. Systems should be flushed with nitrogen and evacuated before charging.

  Insulation - Insulate refrigerant lines in accordance with manufacturers recommendations. (Liquid and suction lines should not be in contact with each other.)

  Capacity Control - Double suction or discharge risers and oil traps may be required on split systems to insure oil return at low loads.

- Protect refrigerant lines which pass through masonary work. Seal all openings in masonary work.
- Filters, driers, strainers, sight glass, solenoid valves, and service valves.
- Leak testing.
- Vibration isolation.
- Supports and hangers.

5.6 Refrigerant Charge

Even if the equipment seems to be operating satisfactorily, the ability of the unit to provide cooling or heating on a given day does not verify that the charge is correct. The ability of the equipment to operate with an incorrect charge varies with the mode of operation and the ambient conditions. For this reason, the charge must be weighed in, or verified, by using pressure-temperature charts. It is possible for an incorrect charge to go unnoticed for a considerable time because the equipment has some tolerance for variations in refrigerant charge. However, efficiency and equipment life will be reduced.

5.7 Condensate Disposal

The condensate that forms on indoor or outdoor coils must be collected and disposed of. Condensate should not be allowed to stain either the exterior or the interior of the structure, or to accumulate in pools.

- Drains should be properly trapped and pitched to insure flow under all operating conditions.
- Drain lines should be as large or larger than the fitting on the condensate pan.
- Drain lines must be protected against freezing.
- Install an overflow pan equipped with a separate drain under the condensate drain pan if structural or cosmetic damage can be caused by a clogged drain. (See Figure 5.7).

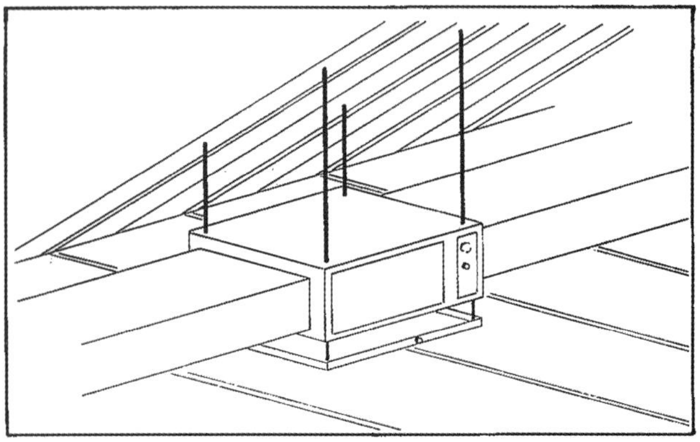

Figure 5.7 Install Overflow Drain Pan

5.8 Sound and Vibration

Sound and vibration can be transmitted through walls, floors and ceilings, or it can be carried to the occupied space by the air distribution system. It is also possible for the air distribution system to generate noise at fittings, dampers, grills or other restrictions.

- Massive walls that are thoroughly sealed will isolate sound from adjacent spaces.
- Use duct liner or lined elbows to attenuate noise carried by the air distribution system.
- Do not install branch volume control dampers near grills and registers.
- Avoid using inefficient duct fittings which produce turbulence.
- Isolate all rotating equipment from the building structure.
- Isolate all ducts, piping and conduit from the equipment and the building structure.

5.9 Power Wiring

Power wiring should be installed in accordance with all codes and regulations and manufacturers recommendations.

- Verify that each piece of equipment is supplied with the correct voltage. (Motors can be damaged by incorrect voltage and electrical resistance heater output varies as the square of the applied voltage.)
- Verify phase balance on three phase circuits.
- Check to see if the electrical service entrance can supply all of the power requirements of the system.

5.10 Control Wiring

Control wiring should be installed in accordance with all codes and regulations and manufacturers instructions.

- Locate outdoor thermostats in a shady protected area.
- Indoor thermostats must be secure and level.
- Locate room thermostats level, and about five feet high, away from internal heat sources and out of direct sun light.
- Check operation and calibration of all controls.

5.11 Air and Water Testing and Balancing

- Adjust fan speed and air side dampers to the specified design CFM.

- Check compressor, fan motor and/or water pump motor AMP draw
- Verify that the air flow through the outdoor and indoor coils are within the limits specified by the manufacturer.
- Adjust air terminals for proper CFM and air pattern.
- Verify that constant air volume regulators are set for the design flow rates.
- Adjust outdoor air, return air, and exhaust air dampers to meet the minimum outdoor CFM requirement.
- Check and adjust water system pressures and flows.

5.12 Accessories

Most manufacturers offer accessories that reduce installation work or increase the effectiveness of the safety and operating controls. Items such as roof curbs, stands, prefabricated plenums, remote control panels, setback controls, status panels, low ambient controls, and anti-short cycling controls are some of the options available.

5.13 Adding a Heat Pump to a Warm Air Furnace or Hydronic Boiler

Warm Air Furnaces

- Verify that the air terminals will provide adequate room air distribution.
- Verify that the fan can handle the increase in static pressure when the heat pump is operating with a wet indoor coil.
- Verify that the air flow through the furnace and the indoor coil is within the limits recommended by the manufacturer.
- Codes, warranties, or problems with condensation and corrosion on the furnace heat exchanger require that the DX coil be mounted downstream from a fossil fuel furnace.
- In some areas, codes allow that the DX coil may be mounted upstream from an electric furnace. (Furnace insulation must be adaquate to prevent condensation and the furnace heating elements must be protected by a high temperature control.)
- Consider reducing the burner capacity on fossil fuel furnaces that are grossly oversized.

Hydronic Boilers

- Verify that the hydronic terminals will provide adequate heat at the water temperatures that are produced when only the heat pump is operating.
- The heat pump should be mounted upstream from the boiler and hot boiler water should not be allowed to circulate through the heat pump.